Oxytocin, Well-Being and Affect Regulation

Eliana Nogueira-Vale

Oxytocin, Well-Being and Affect Regulation

Eliana Nogueira-Vale
University of São Paulo
São Paulo, São Paulo, Brazil

ISBN 978-3-031-59040-5   ISBN 978-3-031-59038-2   (eBook)
https://doi.org/10.1007/978-3-031-59038-2

English translation of the 1st original Portuguese language edition published by Editora Manole, Santana de Parnaíba, 2022

Translation from the Portuguese language edition: "Ocitocina, Bem-Estar e a Regulação do Afeto" by Eliana Nogueira-Vale, © Editora Manole and Eliana Nogueira-Vale 2022. Published by Editora Manole. All Rights Reserved.

© The Editor(s) (if applicable) and The Author(s), under exclusive license to Springer Nature Switzerland AG 2024
This work is subject to copyright. All rights are solely and exclusively licensed by the Publisher, whether the whole or part of the material is concerned, specifically the rights of reprinting, reuse of illustrations, recitation, broadcasting, reproduction on microfilms or in any other physical way, and transmission or information storage and retrieval, electronic adaptation, computer software, or by similar or dissimilar methodology now known or hereafter developed.
The use of general descriptive names, registered names, trademarks, service marks, etc. in this publication does not imply, even in the absence of a specific statement, that such names are exempt from the relevant protective laws and regulations and therefore free for general use.
The publisher, the authors and the editors are safe to assume that the advice and information in this book are believed to be true and accurate at the date of publication. Neither the publisher nor the authors or the editors give a warranty, expressed or implied, with respect to the material contained herein or for any errors or omissions that may have been made. The publisher remains neutral with regard to jurisdictional claims in published maps and institutional affiliations.

This Springer imprint is published by the registered company Springer Nature Switzerland AG
The registered company address is: Gewerbestrasse 11, 6330 Cham, Switzerland

Paper in this product is recyclable.

Dedicated to the memory of Jaak Panksepp (1943–2017)

Foreword to the Brazilian Edition

This book, the result of two decades of dedication to the study and research of neuropeptide *oxytocin*, emphasizing its links with symptoms and syndromes of anxiety and depression, is a pioneer among us.

These two psychiatric disorders have neurobiological aspects, affective and emotional aspects, and behavioral aspects, combining aspects of body and mind of the person, being thus subjects of affective neuroscience.

The text will present a historical review of this neuropeptide, its many functions, structures and neural nets, which compose the oxytocinergic system, which implies that its production by the body is directly related to the behavior of attachment and its relationships with brain development and affective regulation, providing a basic understanding to assist the psychotherapy of anxiety disorders.

It is a book for students interested in the integration of psychology and neuroscience as well as professionals who want to improve and better understand their practice.

One of the most interesting areas of research today is the relationship and impact of neuroscience on psychotherapy. The new imaging technologies, brain computed tomography, and brain magnetic resonance have allowed us to study the structure and better understand the brain function.

Other fields of study have developed, which also contributed to the dialogue between neurosciences and psychotherapy: neurobiology, cognitive sciences, developmental psychology and psychiatry, and ethology.

The discovery of brain plasticity, with the steady renewal and pruning of synapses, revealed a brain in constant formation and adaptation to reality. Its structures and connections are always changing, and new interconnections are being formed.

The discussions about the mind-body relationship, Bowlby's studies on attachment, the importance of early childhood experiences, the formation of secure bonds and their role in personality formation, and the importance of a coherent narrative that gives meaning to everything that happens have changed our view of psychotherapy. Studies on memory formation, the amygdala–hippocampus relationship, the diverse types of memories, the impact of traumatic events, the windows for modifying these memories, and the role of biological interventions integrating different brain areas and forming new connections contributed to strengthen and improve psychotherapy.

Progress has also been made over the importance of self-regulation and its role in improving various mental disorders. Self-regulation involves a coordinated flow of energy and information among the main brain systems. New therapies have sought to improve this regulation through techniques that promote flexibility, such as meditation practices or mindfulness. The development of Positive Psychology included practices to promote happiness and moral behavior, and to better handle self-criticism and shame, through studies on compassion, self-compassion, gratitude, and other related topics. The comprehension of the neural systems involved in negative and positive emotions has progressed and facilitated this integration.

The discovery of neuropeptides, and especially *oxytocin* and its role on psychological functions, emotions, mood, behavior, breastfeeding, childbirth, and strengthening of human bonds and affiliative behavior, brought new perspectives.

Associate Professor of the Faculty of Medicine and the Institute of Psychology at the University of São Paulo
Sao Paulo, Sao Paulo, Brazil

Francisco Lotufo Neto

Preface to the English Edition

This text was originally written and published in Brazilian Portuguese, aiming at the Brazilian public, and was mainly based on the studies carried out for my doctoral thesis, "Oxytocin, Attachment, and Sleep: A Study with People with Generalized Anxiety Disease."

For the English version, dedicated to an international public, certain adaptations appeared to be necessary, which included the withdrawal of the material that reflects solely Brazilian reality and history, and its substitution for a more global view of facts. This meant to enhance the events that took place in the United States, as it was the home for many relevant scientific developments and events related to the world history of mental health, except for psychoanalysis, originated in Europe.

We have also updated the text according to the most recent bibliographical data.

At the end, the reader will find two extra chapters which were not in the original version, a suggestion from my Editor, promptly accepted by me, as it enriches the original text.

University of São Paulo Eliana Nogueira-Vale
São Paulo, São Paulo, Brazil
January 2024

Preface to the English Edition

Acknowledgments

I would like to thank my Editor Bruno Fiuza, for his attentive and kind supervision through the entire process of editing the English version of this book, promptly answering my questions and clarifying doubts, as well as giving many important suggestions in the adaptation of the book for an international version.

Francisco Lotufo Neto, for giving permission to include his foreword of the Brazilian edition in the English version of the text.

Maria Helena Vargas, for carefully preparing this manuscript with scientific expertise, adapting the Brazilian version to Springer's format requirements.

Coaraci Nogueira do Vale, for double-checking the English version.

Ciro Rodrigues de Araujo, for adapting the illustrations to the English version.

And the Springer's team, for supervising the final English version of this book.

Contents

1 Introduction... 1
2 Affective Neuroscience and the Discovery of Oxytocin............ 9
3 The Nervous System and the Affective Neural Systems........... 19
4 Oxytocin.. 37
5 Attachment 51
6 Brain Development.. 65
7 Development and Affective Regulation........................ 75
8 Therapeutic Choices... 85
9 A Conversation Between Affective Neuroscience
  and Psychoanalysis.. 99

Index... 109

About the Author

Eliana Nogueira-Vale is a Brazilian clinical psychologist and neuroscientist with a master's degree in clinical psychology and a doctorate in Neuroscience and Behavior from the Institute of Psychology (IP) at the University of São Paulo (USP) (2019). She is the Main Researcher in Pilot Project: Natural Stimulation of Oxytocin and Affect Regulation: an investigation with middle-aged women in the city of São Paulo at Federal University of São Paulo (2010) (unpublished), and Main Researcher of the Thesis: Oxytocin, Attachment and Sleep in People with Generalized Anxiety Disorder (USP 2019). She was a Member of the Mind-Body Group of the Brazilian Society of Psychoanalysis of São Paulo (2005–2007); of The International Neuropsychoanalysis Society (since 2001); of the Anesthesiology Pain Control Group, Faculty of Medicine of USP (from 2007 to 2015) and of the SuCor Laboratory—Subject and Body, Clinical Psychology of IP USP (2015–2019). First Author of Oxytocin & well-being as promoters of affect regulation and homeostasis: a neuroscientific review. Psico, Porto Alegre, v. 51, n. 2, p. 1–12, abr.-jun. 2020. Books: *[Os rumos da psicanálise no Brasil: um estudo sobre a transmissão psicanalítica]. The pathways of psychoanalysis in Brazil: a study on psychoanalytic transmission* (Ed. Escuta 2003), of *[Ocitocina, bem-estar e regulação do afeto] Oxytocin, well being and affect regulation* (Ed. Manole 2022), and several chapters in other books, as well as well as scientific articles in journals. Private Clinical Practice (since 1975).

Chapter 1
Introduction

This book was written at a traumatic time in our lives. We were amid the COVID-19 pandemic, a period when we lived an unprecedented circumstance—"I never imagined that we could go through a situation like this!", claimed people—, worsened by the almost no scientific knowledge about the disease; by the literal threat to our lives, the subversion on personal and professional routines restricting our circulation in public spaces and the dismantling of the organization of school routine, of work, economy, and culture as we used to live in. We found ourselves in harsh isolation, unable to predict what face the world would show next. The pandemic deprived us of physical and face-to-face contact for a prolonged period, which affected personal relationships among people.

Coincidentally, *oxytocin* (OT), the subject of this book, plays a fundamental role in the organization of our social life and in the creation of affective bonds among people (Heinrichs et al. 2009).

OT is a brain hormone, typical of mammals, including humans. It curiously exhibits a trace of sexual dimorphism: it is produced in greater quantity in the brains of females (and women) than brains of males (and men), differentiating male brains from female, which would cause social consequences, as will be explained in this book (Gao et al. 2016).

OT began to be publicized by the media and became popular, being called, simplistically, the "love hormone," which gave rise to a series of sensationalist misunderstandings and shallow scientific information. We hope to help in clarifying such a distortion with this book.

The history of OT begins in 1900, when the English scientist Sir Henry Dale discovered that the hormone accelerated uterine contractions during childbirth and stimulated the descent of milk (Dale 1906). Since then, OT has been used in hospitals for the alleged purposes, but also for the containment of uterine hemorrhages during childbirth. Its use extended to veterinary practice with the same purpose, becoming a classic in medicine and in veterinary till today.

© The Author(s), under exclusive license to Springer Nature Switzerland AG 2024
E. Nogueira-Vale, *Oxytocin, Well-Being and Affect Regulation*,
https://doi.org/10.1007/978-3-031-59038-2_1

Before proceeding with the history of OT, which has not stopped over the years, let us contextualize the scientific moment when these discoveries were made.

The end of the 1890s was marked by the beginning of scientific discoveries about the nervous system (NS) (Ehrlich and Cassidy 2021).

The main discoveries were as follows: (i) identification of neurons as basic units of the architecture of the NS; (ii) communication between them through nerve impulses; (iii) characterization of nerve impulses as transductions of neural messages between the body and the brain, to maintain the homeostasis of the organism; and, finally, (iv) understanding the process of neurotransmission of messages carried out by neurons, with the help of neuroactive substances (Venkataramani 2010).

At the end of the 1890s, the Spanish Santiago Ramón y Cajal created the neuronal theory, articulating all the discoveries made so far about the NS (López-Muñoz et al. 2006). In 1906, he received the Nobel Prize for his contribution and became known as "the father of modern neurosciences" (Venkataramani 2010).

To recognize a father of neurosciences is also a sign that we entered a new scientific era, and the discovery of OT will occur within this new neuroscientific framework.

By the end of the nineteenth century, neurons were gradually being identified as basic elements of the nervous system (NS), as well as their neuroanatomical (soma or cell body, dendrites, and axon) and functional characteristics (they would be units that transmitted messages between the brain and the body and would communicate with each other through synapses).

And how did the messages transmitted through neurons operate? They moved through the neural networks that formed the NS, carried by neuroactive substances, until reaching a target organ, where the message was delivered, and could eventually turn into action.

In the first decade of the twentieth century, John Langley created the term "receptor molecules" to identify access channels in the membrane of the target neuron of the message, in a "key-and-lock" type mechanism (Maehle 2004). Usually, a receptor gives access to one type of neurotransmitter, the one that fits its lock.

Twenty years later, in England, in the laboratory of Sir Henry Dale (the same scientist who had discovered OT), with the collaboration of qualified foreign scientists, two initial neurotransmitters were discovered: the *substance P* (1931) (Euler and Gaddum 1931) and *acetylcholine* (1936) (Finger 2010).

If we review the history of neuroscientific discoveries more closely, one realizes that they become viable as new techniques are made available.

Around 1970, thanks to modern technologies and discoveries, there was a boom of new research. One of the most innovative and exciting discoveries was the identification of neurotransmitters that were named *neuropeptides* (Russo 2017). The revolutionary aspect of neuropeptides was that they controlled *basic psychological functions, discrete emotions, and mood states*, with direct consequences on *behavior* (Panksepp 1993). In other words, they constituted a link between *mind, brain, and body*, putting an end to centuries of Cartesian mind-body division. Importantly, OT is part of this revolutionary group of neuropeptides, and there is a wide range of possibilities to be explored.

The unraveling of the neuroanatomy and functioning of the *basic affective neural systems* was only possible thanks to the development of more powerful neuroimaging, innovative technologies for clinical exams, and many detailed works of investigation and dissection in animal brains (Panksepp 1998).

These animal studies served as a prediction or parameter for human behavior (Panksepp 2011), by means of an approach called *comparative neuroscience*, and its merit will become apparent when one examines the typical characteristics of mammalian brains.

Studies of OT with humans did not begin before 2000, due to the fact that the techniques known at that time were invasive (Kendrick et al. 2017). We are, therefore, dealing with a type of very recent scientific research, and in full development. Currently, we know that OT plays an essential role in numerous areas, the details of which will be presented throughout this book: social behaviors in general, attachment, regulation of affect, homeostasis, analgesic qualities, healing and regenerating of body tissues, production of states of calm, well-being, trust, security, achievement, and happiness, just to mention a few (Uvnas-Moberg 2011).

From now on, we will deal with OT in the human species. When we refer to lower mammals, this will be specified.

As we have seen, OT is produced both in the brains of males and females, but because the scale of production is greater in women, it became known, somewhat inaccurately, as a female hormone (see Gao et al. 2016).

Women exhibit a *significant increase in OT production at the end of pregnancy and during labor* (Uvnäs-Moberg et al. 2019). This pattern will be maintained in the mother's brain as well as in the baby's, approximately during the first 3 years of the child's life; it is currently known that fathers can also produce OT in physical contact with their children (Scatliffe et al. 2019). The greater availability of OT in the mother's and baby's brains will favor the development of a special relationship between them, mediated by processes of affective-emotional *attachment* and *mothering* (the process of delivering maternal care to the baby).

Attachment and maternal care, in turn, directly contribute to the *development of brain structures* in the baby—they help to *sculpt* the brain (Endevelt-Shapira and Feldman 2023). It should be noted how a psychological phenomenon (affective-emotional attachment) has a direct influence on neurobiological development (brain development).

The human baby will need maternal care for a critical period, much longer than most lower mammals, as they are born in a pronounced state of neurobiological prematurity and dependence on the caretaker, usually the mother (Somel et al. 2009). Lower mammals are born with a much greater readiness for life outside the womb, with a wide instinctive and automatic repertoire phylogenetically established, characteristic of the species, and genetically inherited. A puppy, for instance, can walk by itself within a few days, and a foal can stand up shortly after birth. Even though there are similarities between mammals, there are also many specific aspects to each species, and caution is needed when establishing comparisons between distinct species.

In humans, thanks to neural plasticity, the brain, body, and motor development will occur much more slowly and in a personal way, allowing the environment to cause stronger influences in the outcome of humans, in opposition to lower animals, which are more instinctive.

The child's autonomy and its attachment dynamics will be modulated by the style of maternal care (Holmes 1997), and greatly influenced by the cultural norms of their environment. As we are speaking animals, the mother's words will introduce the baby into the world of language before birth; it will hear the mother's words from the womb. After birth, she will interpret for the baby, what is happening around at that moment, verbally, prosodically, with the melodic line of the mother's voice, intonation, and rhythm. This occurs long before the child develops symbolic capacity and speech. At this stage, the mother's speech appears in the form of a melodic flow, and the baby emotionally registers this discourse in his brain (Bornstein et al. 2012). As the child is not a *"tabula rasa,"* his genotype will also play a role in this process.

The fact of the mother's voice, with its authentic intonation and modulation, its melody, is essential for the formation and strengthening of a true bond between mother and baby. And the model for it, you may have recognized, is that of *maternal attachment*, the matrix of affective relationships with the significant other(s). Still as a symbolic reference, let us remember that attachment is an evolutionary phenomenon responsible for providing *security* to the baby, and is essential for their *survival*.

The simultaneous production of OT and the behavior of attachment contribute to the creation of an intense loving bond between mother and baby, and this will be the vehicle for a process of affective regulation between both (Galbally 2011).

Affective regulation depends on the mother's ability (or inability) to meet the affective and physical demands of her baby, addressing the baby when he needs her; calming when he is scared, angry, in pain or hungry; dressing him; and cleaning him when necessary. On the other hand, stressful situations produce such an excitatory intensity that the baby, due to his premature neurobiological mind, goes into despair and panic if not rescued. The mother will need to put a lot of effort to maximize her emotional and cognitive abilities to identify with her child's suffering and to soothe the child until he gradually gains emotional autonomy. It is worth adding that, at the beginning of life, the baby is introduced to so many things that he does not yet know. These novelties, whether positive or negative, can be stressful, due to their low excitatory threshold. In addition, no matter how dedicated the mother is, she is also subject to her own stressors and distractions and will not always be there to meet the baby's needs. In other words, the baby will also have to deal with the frustrations caused by the delay in having his desires satisfied, and frustration will be part of the development process. In a proper proportion, the mother is expected to be "good enough," according to the concept of psychoanalyst Donald Winnicott (1896–1971) (Winnicott 1978), that is, she is not perfect.

The neurodynamics of attachment is characterized by a strong and constant *need for physical proximity* between mother and baby, for *eye contact* and *physical touch*, for smelling each other, and for sound exchanges and babbling, which contribute to moments of great happiness, fulfillment, trust, and well-being between both, as well as their affective regulation (Bahn 2022).

Some authors prefer to use the term *homeostasis*, created by the physiologist Walter Cannon (1871–1945), to designate a characteristic of the most complex living organisms, which is the ability to restore their internal stability after they suffer a disruption caused by excitatory situations. For Cannon, homeostasis is maintained within strict limits and occurs automatically. Cannon, however, only considers homeostasis as a *physiological* regulation (Davies 2016).

More recently, some authors have adopted the concept of affect regulation to refer to the balance between brain, mental and body aspects, plus environmental factors, for an affective regulation. Authors, who follow this concept, place the primacy of the regulation in the affective component (Fonagy et al. 2005; Schore 2016).

Affects triggered in brain neural systems by external and internal stimuli (and not from "psychological creations") are an important evolutionary alert in terms of the organisms' survival, being able to vary in intensity, affects warn whether everything is okay with the person or if she/he is in danger, threatened, or running a risk upon their life.

When OT is produced in the body, an affective state of well-being and comfort tends to settle in the person. It would be the equivalent of the reunion of a child with their mother. In other words, only after calming down and restoring emotional balance can the child be properly fed, relax, and fall asleep.

This model of affect regulation also applies to other age groups. Anyone who has tried to eat after a strong annoyance or affective problem knows well that one feels no hunger, cannot swallow the food, and, if they do, there is a chance that they will feel sick.

OT also has an important function in restoring the organism of adults after situations of stress, inflammation, or extreme physical/mental exhaustion in professional life, the so-called syndrome of *burnout* (Takayanagi and Onaka 2021).

In situations of bodily harm, OT helps to accelerate the restoration of damaged body tissues and the healing of injuries (Uvnäs-Moberg et al. 2019). It has analgesic effects, attenuating the sensation of pain. This can happen naturally or be boosted by the supportive presence of another human being. The presence of a significant other, who comforts the sick person during an invasive procedure, generally reduces their pain and suffering.

Because it covers such a broad and important scope of effects related to the production of OT, this book is addressed to students of psychology, medicine, and other health professionals, but also to the lay reader interested in life sciences. Thus, I will avoid the excessive use of technical and academic terms, to facilitate the understanding of the text.

Throughout this book, I will tell how knowledge about OT unfolded, from its discovery, at the beginning of the twentieth century, until today. I will also talk about its fundamental implications for the mental and physical health of people, including in the treatment of psychiatric disorders, although research in this area is still inconclusive. However, the most recent developments in research will be mentioned.

Since we will refer to neuroanatomic structures of the human body, it is advisable to have an atlas of the human body at hand, to follow and visualize their descriptions.

I hope you enjoy this book. I also hope that we can socially reinvent ourselves, in this post-pandemic scenario, where our affective contacts were subjected to the laws of social distancing.

References

Bahn GH (2022) Understanding of holding environment through the trajectory of Donald Woods Winnicott. J Korean Acad Child Adolesc Psychiatry 33(4):84–90. https://doi.org/10.5765/jkacap.220022

Bornstein MH, Suwalsky JT, Breakstone DA (2012) Emotional relationships between mothers and infants: knowns, unknowns, and unknown unknowns. Dev Psychopathol 24(1):113–123. https://doi.org/10.1017/S0954579411000708. PMID: 22292998; PMCID: PMC3426791

Dale HH (1906) On some physiological actions of ergot. J Physiol 34(3):163–206

Davies KJ (2016) Adaptive homeostasis. Mol Aspects Med 49:1–7. https://doi.org/10.1016/j.mam.2016.04.007

Ehrlich KB, Cassidy J (2021) Early attachment and later physical health. In Thompson RA, Simpson JA, Berlin LJ (eds) Attachment: The fundamental questions. The Guilford Press, pp 204–210

Endevelt-Shapira Y, Feldman R (2023) Mother–infant brain-to-brain synchrony patterns reflect caregiving profiles. Biology 12(2):284. https://doi.org/10.3390/biology12020284

Euler USV, Gaddum JH (1931) An unidentified depressor substance in certain tissue extracts. J Physiol 72(1):74–87. https://doi.org/10.1113/jphysiol.1931.sp002763

Finger S (2010) Otto Loewi and Henry Dale: the discovery of neurotransmitters. In: Minds behind the brain: a history of the pioneers and their discoveries. Oxford Academic, New York. https://doi.org/10.1093/acprof:oso/9780195181821.003.0016. Online edn, Accessed 31 Oct 2023

Fonagy P, Gergely G, Jurist E, Target M (2005) Affect regulation, mentalization, and the development of self. Routledge

Galbally M, Lewis AJ, Ijzendoorn Mv, Permezel M (2011) The role of oxytocin in mother-infant relations: a systematic review of human studies. Harv Rev Psychiatry 19(1):1–14. https://doi.org/10.3109/10673229.2011.549771. PMID: 21250892

Gao S, Becker B, Luo L, Geng Y, Zhao W, Yin Y, Hu J, Gao Z, Gong Q, Hurlemann R, Yao D, Kendrick KM (2016) Oxytocin, the peptide that bonds the sexes also divides them. Proc Natl Acad Sci U S A 113(27):7650–7654. https://doi.org/10.1073/pnas.1602620113. Epub 2016 Jun 20. PMID: 27325780; PMCID: PMC4941426

Heinrichs M, von Dawans B, Domes G (2009) Oxytocin, vasopressin, and human social behavior. Neuroendocrinology 30(4):548–557

Holmes J (1997) Attachment, autonomy, intimacy: Some clinical implications of attachment theory. Br J Med Psychol 70(3):231–248

Kendrick KM, Guastella AJ, Becker B (2017) Overview of human oxytocin research. https://doi.org/10.1007/7854_2017

López-Muñoz F, Boya J, Alamo C (2006) Neuron theory, the cornerstone of neuroscience, on the centenary of the Nobel Prize award to Santiago Ramón y Cajal. Brain Res Bull 70(4-6):391–405. https://doi.org/10.1016/j.brainresbull.2006.07.010. Epub 2006 Aug 14. PMID: 17027775

Maehle A-H (2004) "Receptive Substances": John Newport Langley (1852–1925) and his path to a receptor theory of drug action. England. Cambridge University Press

Panksepp J (1993) Neurochemical control of moods and emotions: Amino acids to neuropeptides. In: Lewis M, Haviland JM (eds) Handbook of emotions. The Guilford Press, pp 87–107

Panksepp J (1998) Affective neuroscience: The foundations of human and animal emotions. Oxford University Press

References

Panksepp J (2011) The basic emotional circuits of mammalian brains: Do animals have affective lives? International Publishing AG 2017 Curr Topics Behav Neurosci Biobehav Rev 35(9):1791–804. https://doi.org/10.1016/j.neubiorev.2011.08.003. Epub 2011 Aug 19. PMID: 21872619

Russo AF (2017) Overview of Neuropeptides: Awakening the Senses? Headache 57 Suppl 2(Suppl 2):37–46. https://doi.org/10.1111/head.13084. PMID: 28485842; PMCID: PMC5424629

Scatliffe N, Casavant S, Vittner D, Cong X (2019) Oxytocin and early parent-infant interactions: A systematic review. Int J Nurs Sci 12;6(4):445–453. https://doi.org/10.1016/j.ijnss.2019.09.009. PMID: 31728399; PMCID: PMC6838998

Schore AN (2016) Affect regulation and the origin of the self: the neurobiology of emotional development, 1st edn. Routledge

Somel M, Franz H, Yan Z, Lorenc A, Guo S, Giger T et al (2009) Transcriptional neoteny in the human brain. Proc Natl Acad Sci USA 106(14):5743–5748. https://doi.org/10.1073/pnas.0900544106

Takayanagi Y, Onaka T (2021) Roles of oxytocin in stress responses, allostasis and resilience. Int J Mol Sci 23(1):150. https://doi.org/10.3390/ijms23010150

Uvnas-Moberg (2011) The oxytocin factor: tapping the hormone of calm, love, and healing. Peter and Martin. ISBN:9781905177639

Uvnäs-Moberg K, Ekström-Bergström A, Berg M et al (2019) Maternal plasma levels of oxytocin during physiological childbirth – a systematic review with implications for uterine contractions and central actions of oxytocin. BMC Pregnancy Childbirth 19:285. https://doi.org/10.1186/s12884-019-2365-9

Venkataramani PV (2010) Santiago Ramón y Cajal: father of neurosciences. Reson 15:968–976. https://doi.org/10.1007/s12045-010-0113-6

Winnicott DW (1978) Playing and reality. Routledge

Chapter 2
Affective Neuroscience and the Discovery of Oxytocin

Neuroscience: From Antiquity to the Present Day

How and When Did Neuroscience Emerge?

There are indications of precursor investigations in neuroscience since the ancient times of Egypt and Greece. The first known manuscript on the subject (The Edwin Smith Surgical Papyrus) dates from 1700 BC (UCLMR 2018). Only in 1962, however, was the term "neuroscience" officially created by Francis Otto Schmitt, who, in the occasion, defined it as an *"interdisciplinary research program, bringing together the various physical, biological and neural sciences ... to attack a single goal, to understand the connections between mind, brain, and behavior"* (Adelman 1910, p. 8).

It is striking that *affective neuroscience*, an important area of neuroscience, was left out of this definition: the one that deals with *affective-emotional phenomena*. However, this is not surprising. For much of the twentieth century, most scientists showed contempt for the study of affects and feelings as scientific phenomena. Some thought they were not susceptible to objective identification/quantification; others, that they would be phenomena outside the scientific field, and others, still, that these did not even exist (LeDoux 1999). Affective neuroscience came to rescue affects and emotions for a more integrative and scientific conception of the human being.

Precursor Neuroscientists

Among the precursors of modern affective neuroscience, we can highlight Darwin and his Theory of Evolution (1872) (Darwin 1859); Broca and the discovery of the "language center" in the brain (1872) (Dalgleish et al. 2009); Exner and his work in comparative physiology (1894) (Verstraten et al. 2015); and his friend Freud and the creation of psychoanalysis (1895) (Freud 1953). The theories of these precursors, elaborated in parallel, but independently, have in common the fact of being based on models of emotional networks, anticipating future theories.

Pioneer Neuroscientists

In the United States, the James-Lange theory (1894) (Cannon 1927), which proposed that emotional experiences were based on bodily events, was considered the first relevant work to be published in the field of affective neuroscience. We should also remember Santiago Ramón y Cajal, already mentioned, who received the Nobel Prize for his Neuronal Theory and the nickname of "Father of Neuroscience."

These works were followed by others, such as those by Cannon (1927), MacLean (1990), Panksepp (1998), LeDoux (1999), and Damásio (2008), with important contributions on the role of certain *brain structures* in affects such as the limbic lobe, the amygdaloid complex, and the prefrontal cortex.

It is beyond the scope of this chapter to provide a broader exposition on the work of these authors, but we will emphasize some relevant aspects for this book.

- Cannon coined the term *homeostasis* to designate the body's state of balance, and the rupture of it by stressing events, activating the hypothalamus-pituitary-adrenal (HPA) axis.
- MacLean created the well-known *triune brain theory*, of evolutionary conception, and his disciple Panksepp developed a theory about the *affective neuroscience*, which includes the identification and description of basic neural affective-emotional systems in the mammalian brain.
- LeDoux, among many relevant works, conducted an important study on the *fear neural circuit*.
- Damásio, while studying the clinical case of Phyneas Gage, suggested that there is no socially adequate psychic and behavioral functioning if the affective aspects (emotions and feelings) are not integrated to the cognitive ones (logical reasoning). The impressive report about Phyneas Gage, who had his brain perforated by a metal rod, and, because of the brain injury, stopped associating his emotions and feelings with cognitive processes, which resulted in the emergence of an inability to empathize, and he could no longer lead a normal social life.

Neuroscience and Neuropeptides in the Twentieth and Twenty-First Century

In the early 1950s, a new class of *neuroactive substances* was discovered—the *neuropeptides*, whose name was later coined in 1971, by the Dutch neuroscientist David de Wied (1925–2004) (De Wied et al. 1993).

Neuropeptides were discovered in an occasion when a group of scientists was researching the properties of hypothalamic hormones, focusing their neurobiological properties. This beginning of research was quite arduous, as the established objectives took about 20 years to be achieved, because of the lack of technological tools.

Research took a leap in the 1970s, when it was discovered the relationship between the synthesis of *neuropeptides* and their determining role in the emergence of *affects*. This caused a revolution in the comprehension of the human being: neuropeptides circulate neuro-affective information between the *endocrine, behavioral,* and *psycho-affective systems* (Panksepp 2003), in addition to influencing the *immune responses* and the *autonomic* and *vegetative nervous system* (Klavdieva 1996).

The neuroscientist Candace Pert, internationally known for having discovered the *opioid receptor* in the brain when she was still an undergraduate student, and for her subsequent research on neuropeptides and their receptors, called them "*psychosomatic network*," because, in fact, they are responsible for the articulation among the "*mind-body*" systems (sic) (Schwartz 2013).

According to Pert:

> Clearly, the conceptual division between the sciences of immunology, endocrinology, and psychology/neuroscience is a historical artifact: the existence of a communicating network of neuropeptides and their receptors provide a link among the body's cellular defense and repair mechanisms, glands, and brain. (1985, p 824s)

The comprehension about the properties of neuropeptides also made it possible to understand a new way of relating mind, brain, body, and behavior. Jaak Panksepp (1998) proposed a model of "triangulation *[between] affective experience, behavioral and bodily changes, and the operation of neural circuits, concurrently.*" The author adds that certain emotional states are "*neurologically quite primitive, since they appear to be triggered by the arousal of various emotional subcortical circuits*" (p. 34). This last statement is relevant for clinical care, as it clarifies how the same person can act in a mature way at a certain moment, and in a totally primitive, childish, and unconscious way at another, depending on when and which circuits will be activated. Several contemporary biological psychiatrists understand that there is a potential role for neuropeptides in the treatment of psychiatric diseases, although research in this area is still in its early stages (MacDonald and Feifel 2013).

Human research on neuropeptides, at the beginning of the twenty-first century, confirmed that many of their psychophysiological and behavioral aspects were analogous to those already observed in mammalian animal models, hence the importance of comparative studies between men and less evolved mammals, especially those that have brains homologous to humans.

The advancement in understanding neuropeptides was slow and lasted many years; it continued thanks to the parallel development of certain technologies:

- New laboratory techniques, such as *radioimmunoassay*, discovered in 1959 (Panksepp 1998).
- Advances in neuroimaging, with increasingly expanded and defined resolutions of very small neurobiological elements. Functional magnetic resonance imaging (fMRI), for example, allows a person's brain to be observed while he performs a certain task, which allows to identify which brain regions are activated in this circumstance (Functional Magnetic Resonance, accessed in 2021).
- The creation and expansion of a global network of virtual communications among scientists came to facilitate contact, the storage of enormous amounts of data [*big data*], the processing of information, its real-time transmission among neuroscientists, and the integration of an increasingly complex study (Pechura and Martin 1991).
- More recently, neuroscientific progress in molecular genetics and genomics has helped to unravel the role of variants and mutations of genes associated with major psychiatric disorders, contributing to the creation of large online databases, thus expanding the horizons of biological psychiatry, and facilitating the access to content in real time (Konopka and Geschwind 2010).

The History of the Neuropeptide OT in Neuroscience

Discovery of Peripheral OT

OT, the subject of this study, is one of the neuropeptides with the most complex action in the mind-body relationship. At the time of its discovery, however, we knew nothing about neuropeptides and their functions.

OT was identified and named by the British physician, physiologist, zoologist, and Nobel laureate Sir Henry Dale in 1909, who, while conducting research for the Wellcome laboratory in London in the physiological research section, discovered that a certain extract produced by the posterior pituitary gland was able to provoke contractions in the uterus of a pregnant cat (Dale 1909).

In 1910, Ott and Scott published an article in which they reported that the same extract also facilitated the ejection, in pulses, of milk from a lactating woman. From 1911, this extract began to be used by obstetricians to stimulate and increase uterine contractions, with the aim of inducing labor in women (Dalton 2005).

Fifty years later, the American biochemist Vincent Du Vigneaud sequenced and synthesized OT, a pioneer work in sequencing a neuropeptide hormone. Du Vigneaud won for this work, the Nobel Prize, in 1955 (Magon and Kalra 2011).

In summary, from then until the present moment, synthetic OT has had its role consecrated in clinical medicine for the induction of labor, in the containment of uterine hemorrhages, and in the stimulation of milk letdown, in obstetrics and in veterinary medicine. These are called *peripheral* effects of OT, as they occur outside the central nervous system.

Discoveries About Central OT in the Twentieth and Twenty-First Centuries

The central effects of OT, which occur due to its action within the central nervous system, began to be unraveled in the 1970s, as already said. At that time, it was discovered that its production was related to *positive mood states*. Simultaneously, another line of studies began to study the relationships between OT and the *activation of the maternal care system*, and the *bond between mother and offspring* (Carter 2003). These data came from animal research, as human research had not yet begun.

At the end of the 1980s, a group of scientists from the Karolinska Institute, in Sweden, began to study the functions of neuropeptides in humans. Among these pioneers, the neuroscientist Kerstin Uvnäs-Moberg stood out, initiating a long and consistent study on OT and other peptides in the mother-baby relationship, which continues to this day (Uvnäs-Moberg 1996). Another important aspect of her work associates the production of OT with the *state of well-being* (Uvnäs-Moberg and Carter 1998). Possibly, the pre-internet culture of that time, combined with the language barrier, played a role as an obstacle to an earlier international dissemination of these studies.

Meanwhile, in the United States, in the year 1987, Pitman et al. published an article about a pioneering clinical trial, with a double-blind randomized sample and a control group, consisting of war veterans who had post-traumatic stress disorder (PTSD). In this trial, the pioneering technique of administering *intranasal OT* was used. Despite this early start, research with humans only escalated after 2000.

In 1996 and 1997, two innovative collections on the influences of OT production in *maternal, sexual, social,* and *affiliative behaviors*, presenting animal studies carried out by heavyweight researchers, were published by the *Annals of The New York Academy of Sciences* (AAAS) (Pederson et al. 1992).

In 1998, Oxford University Press published the monumental text by Jaak Panksepp on affective neuroscience, with his conceptions on the foundations of animal and human emotions, together with numerous and detailed maps of *discrete neural circuits*, derived from his studies with mammals, especially the rat. These circuits, when activated by electrical or chemical stimulation, elicited the corresponding basic affect (Panksepp 1998).

In the same year, a *special* issue of the journal *Psychoneuroendocrinology: Is there a neurobiology of love?* was published, which studied the relationships of OT with love. It was edited by the invited neuroscientists C. Sue Carter and Kerstin Uvnäs-Moberg, whose work, Carter in the United States, and Uvnäs-Moberg in Sweden, followed parallel paths in the investigation of neuropeptide OT. They attributed to OT associated to social behaviors, emotional states of calm, anti-stress, and positive feelings. However, another 6 years passed before Uvnäs-Moberg began to use the term *well-being* to designate a pleasant and diffuse background state, elicited by tactile stimulation of massage, which, it was then presumed, was related to the production of OT. Until that point, she had used the terms *non-noxious, anti-stress,* and *self-soothing* to refer to this state. Interestingly, in the first two examples, she defined well-being by its opposite.

We know how difficult it is to name affective states in sciences, especially the positive ones. Scientific studies usually prefer the study of negative states, such as stress, pain, suffering, rather than positive states. Positive states are more subtle, less salient. But whether because Uvnäs-Moberg decided to give this turn to designate well-being more directly, or because positive psychology was beginning to gain prestige, the fact is that, from then on, the terminology *well-being* was widely adopted by OT researchers in studies with humans.

As far as we know, the term "well-being" was not used in the pioneering animal literature on OT. However, other anthropomorphic terms associated with OT that could be included in this category were used, such as *love, warmth, sense of ease, positive erotic feelings, playfulness, feelings of security, positive social memories, peaceful coexistence, flirtatiousness, sexual pleasure* and *companionship, orgasm, friendship, sleep, yawning, sedation, intimacy, passion, attachment, calmness, relaxation, homeostasis, regulation,* and *anti-stress*.

This survey, although not exhaustive, may be useful to those who wish to research well-being linked to OT, or build specific scales for these constructs.

Continuing the legacy of MacLean, the Estonian neuroscientist and psychobiologist based in the United States, Jaak Panksepp (1942–2017), continued his research on the relationships between *brain's subcortical systems and basic affects* in mammals, a work to which he dedicated his entire life. In addition, Panksepp also became interested in the *mind-body question* in humans, especially with regard to human consciousness and the *first-person methodology* (this refers to what the speaker knows about himself, in a philosophical dimension) (Chalmers 2017). In this search, and through his contacts with other areas of knowledge, Panksepp came to be interested in *psychoanalysis*.

At the end of the decade of the brain (1990–1999), a significant database on OT research with animals had been accumulated. The growing popularization of electronic means greatly contributed to the quick and direct communication between scientists, including those who traditionally did not dialogue with each other, such as neuroscientists and psychologists.

At this time, Panksepp approached a small study group composed of neuroscientists and psychoanalysts, organized at the New York Psychoanalytic Society by the psychoanalyst Arnold Pfeffer (1915–2002), which resulted in the foundation of *The International Neuro-Psychoanalysis Society (N-PSA)* in 2000, of which Panksepp was co-founder and co-president, along with the South African neurologist and psychoanalyst Mark Solms, until the former's death in April 2017.

Several renowned neuroscientists joined the N-PSA. In 2002, António Damásio, Eric Kandel, Joseph LeDoux, Vilayanur Ramachandran, Oliver Sacks, and Allan N. Schore, in addition to the psychoanalysts Arnold Pfeffer, André Green, Otto

Kernberg, and Marianne Leuzinger-Bohleber were part of the editorial board of the *Journal of Neuro-psychoanalysis*. Later, the Society decided, in a plenary meeting in which this author was present, to change its name, joining the terms "*neuro*" and "*psychoanalysis*" into a single word: *Neuropsychoanalysis*, as it currently appears on its *site:* The Neuropsychoanalysis Association: <society@npsa-association.org>. The work of Eric Kandel, *Psychiatry, psychoanalysis, and the new biology of mind* (Kandel 2005) is a product of this time.

More recently, in an article published in 2012, in a special edition of the magazine *Trends in Cognitive Science*, Panksepp and Solms stated that neuropsychoanalysis seeks "*to understand the human mind, especially as it relates to first person experience. It relates to the essential role of neuroscience in such quests. However, unlike many branches of neuroscience, it positions mind and brain on an equal footing,*" and emphasized "the *prevailing extreme reductionism in neuroscience and biological psychiatry.*"

With it, they advocated for an understanding of mind-body phenomena in an integrated, balanced, and deep way.

In his book *Affective Neuroscience*, 1998, Panksepp initially investigates OT in neuro-affective states of love and sensuality; then, in the field of creating social bonds; and finally, in states of loneliness and helplessness. In other words, research in OT has occurred in stages marked by new discoveries, or by the development of techniques or instruments that allows it to take one more step.

Scientific Works on Central OT in Databases

If we take databases as a reliable record to follow the evolution of studies on central OT, we can say that in the early 2000s, these were still in their infancy. Between 1998 and 2008, we found only two citations in PubMed on clinical trials with humans and the use of OT, except for those related to childbirth, breastfeeding, and obstetrics.

In 2005, Kosfeld et al. conducted, in the United States, the second research with humans that made use of intranasal OT administration [the first had been that of Pitman et al. (1987)] to measure the construct *trust* in humans. Since then, there has been an explosion of research with OT, with emphasis on aspects of *social contact*, and, more recently, from studies of *cognitive aspects*, to *psychiatric disorders* and *molecular studies*.

Status of OT Research in Latin America

In Latin America, except for Argentina, Chile, and Brazil, there is a notable gap in research in central OT.

According to a systematic survey from 1995 to 2020, following PRISMA criteria (Page et al. 2021) in the LILACS database, which is considered the most important index of scientific literature in Latin America and the Caribbean, using as indicators the terms "adult human" (and) "oxytocin," we obtained 359 citations indexed with the term "oxytocin," which, after being filtered by the inclusion and exclusion criteria, amounted to a total of eight articles, of which three were Brazilian, two were Chilean, and three were Argentinian, all of them being review works.

The largest producer of academic works so far has been São Paulo, with a prevalence of works at the University of São Paulo.

Finally, confirming the plurality of OT, we found that academic works in Latin America were carried out in departments as diverse as gerontology, neuroscience and behavior, mental health, pharmaceutical sciences, biomedical sciences, nursing, neuropsychiatry and behavioral sciences, clinical psychology and developmental disorders, and molecular genetics.

References

Adelman G (1910) Chapter 1: Neuroscience before neuroscience WWII to 1969. In: History of neuroscience 1969–1995. The Society for Neuroscience, p 8. https://www.sfn.org/about/history-of-sfn/the-creation-of-neuroscience/~/media/SfN/Images/History%20of%20SfN/pdf/HistoryofSfN.ashx

Cannon WB (1927) The James-Lange theory of emotions: a critical examination and an alternative theory. Am J Psychol 39(1–4):106–124

Carter CS (2003) Biological perspectives on social attachment and bonding. In: Carter CS (ed) Attachment and bonding: a new synthesis. MIT Press, London, pp 85–100

Chalmers D (2017) Blog. In: Facing up the problem of the consciousness (1995) Papers on consciousness

Dale HH (1909) The action of extracts of the pituitary body. Biochem J 4(9):427–447. https://doi.org/10.1042/bj0040427. PMID: 16742120; PMCID: PMC1276314

Dalgleish T, Dunn BD, Mobbs D (2009) Affective neuroscience: past, present, and future. Stearns PN (ed). Emot Rev 1(4):355–368

Dalton LW (2005) The top pharmaceuticals that changed the world: oxytocin. Chem Eng News 83:25

Damasio A (2008) Descartes' error: emotion, reason and the human brain, English edn. Vintage Digital

Darwin C (1859) On the origin of species. Alma Classics, London

de Wied D, Diamant M, Fodor M (1993) Central nervous system effects of the neurohypophyseal hormones and related peptides. Front Neuroendocrinol 14(4):251–302

Freud S (1953–1974) The standard edition of the complete psychological works of Sigmund Freud. Translated from the German under the general editorship of James Strachey. In collaboration with Anna Freud. Assisted by Alix Strachey and Alan Tyson, 24 volumes. Vintage, 1999

Kandel ER (2005) Psychiatry, psychoanalysis, and the new biology of mind. American Psychiatric Pub, Arlington

Klavdieva MM (1996) The history of neuropeptides II. Discovery of hypothalamic neurohormones, elucidation of the primary structure of substance p, and further studies on neurosecretion. Front Neuroendocrinol 17:126–153

Konopka G, Geschwind DH (2010) Human brain evolution: harnessing the genomics (r)evolution to link genes, cognition, and behavior. Neuron 68(2):231–244

References

Kosfeld M, Heinrichs M, Zak PJ, Fischbacher U, Fehr E (2005) Oxytocin increases trust in humans. Nature 435(7042):673–676

LeDoux JE (1999) The emotional brain: the mysterious underpinnings of emotional life. Phoenix Paperbacks, Vermont

Macdonald K, Feifel D (2013) Helping oxytocin deliver: considerations in the development of oxytocin-based therapeutics for brain disorders. Front Neurosci 7:35. https://doi.org/10.3389/fnins.2013.00035

MacLean PD (ed) (1990) The triune brain in evolution: Role in paleocerebral functions. Springer

Magon N, Kalra S (2011) The orgasmic history of oxytocin: love, lust, and labor. Indian J Endocrinol Metab 15(Suppl3):S156–S161

Neuroscience in Ancient Egypt | UCL Researchers in Museums [Internet]. blogs.ucl.ac.uk. Available from: https://blogs.ucl.ac.uk/researchers-in-museums/2018/02/21/neuroscience-in-ancient-egypt/

Ott I, Scott JC (1910) The action of infundibulin upon the mammary secretion. Proc Soc Exp Biol Med 8:48–49

Page MJ, McKenzie JE, Bossuyt PM, Boutron I, Hoffmann TC, Mulrow CD, Shamseer L, Tetzlaff JM, Akl EA, Brennan SE, Chou R, Glanville J, Grimshaw JM, Hróbjartsson A, Lalu MM, Li T, Loder EW, Mayo-Wilson E, McDonald S, McGuinness LA, Stewart LA, Thomas J, Tricco AC, Welch VA, Whiting P, Moher D (2021) The PRISMA 2020 statement: an updated guideline for reporting systematic reviews. Syst Rev 10(1):89

Panksepp J (1998) Affective neuroscience: the foundations of human and animal emotions. Oxford University Press

Panksepp J (2003) Feeling the pain of social loss. Science 302(5643):237–239

Panksepp J, Solms M (2012) What is neuropsychoanalysis? Clinically relevant studies of the minded brain. Trends Cogn Sci 16(1):6–8

Pechura CM, Martin JB (eds) (1991) Institute of Medicine (US) Committee on a National Neural Circuitry Database. Mapping the Brain and Its Functions: Integrating Enabling Technologies into Neuroscience Research. Washington (DC): National Academies Press (US). PMID: 25121208

Pedersen CA, Caldwell JD, Jirikowski GF, Insel TR (eds) (1992) Oxytocin in maternal, sexual, and social behaviors. New York Academy of Sciences

Pitman RK, Orr SP, Forgue DF, de Jong JB, Claiborn JM (1987) Psychophysiologic assessment of posttraumatic stress disorder imagery in Vietnam combat veterans. Arch Gen Psychiatry 44(11): 970–975

Schwartz J (2013) Candace Pert, 67, explorer of the brain, dies. The New York Times

Uvnas-Moberg K, Carter CS (eds) (1998) Psychoneuroendocrinology (special issue) 23(8):749–750. Introduction to volume: is there a neurobiology of love?

Verstraten FAJ, Niehorster DC, van de Grind WA, Wade NJ (2015) (Sigmund Exner's (1887) Einige Beobachtungen uber Bewegungsnachbilder (Some Observations on Movement Aftereffects): An illustrated translation with commentary. 1-Perception 6(5). https://doi.org/10.1177/2041669515593044. License CC BY 3.0593044

Chapter 3
The Nervous System and the Affective Neural Systems

For a better understanding of this book, we will recall some basic concepts of neuroscience: the human nervous system, the brain, cortical and subcortical brain areas, neuronal and glial functioning, synapses, neurotransmitter substances, and basic affective systems.

The Human Nervous System

The human nervous system (NS) is a neural communication network system, with the function of capturing stimuli from the internal and external environments that the organism may consider relevant from the point of view of its *survival* (Mobbs et al. 2015) or the *best choice* for its survival, or to have access to sources of positive stimuli and avoid sources of negative stimuli (De Houwer and Hermans 2001); and then integrate them, process them, and respond to them.

From the point of view of affective valence, environmental stimuli can be classified as *positive* or *negative* (Panksepp 1998), according to the automatic affective reaction they awaken in the person. Positives are those stimuli linked to *satisfaction, well-being, and pleasure*, and the negatives to *pain, discomfort, and displeasure* (De Houwer et al. 2001), which can vary in intensity. We are using here the terms *negative* and *positive* only by their *hedonic value*, and not as synonyms for behavior reinforcers.

The *perception of pain* and its correct *evaluation* are very important factors concerning the *survival and well-being* of the human being. You have probably already seen a *Visual Analog Scale* (VAS), with *emojis* showing different facial expressions, accompanied by numbers ranging from 0 to 10, used by health professionals *to assess the intensity of pain* felt by a patient, with zero equal to no pain, and 10 the greatest pain (Klimek et al. 2017). This type of evaluation, widely used for its practicality, assumes that the perception of pain is a subjective phenomenon, which can only be evaluated from one's own experience, by means of a first-person report (Leal and de Serpa Jr 2013; Schraube 2014), that is, from the person who suffers and speaks.

Currently, however, due to its clinical importance, the questioning about pain is under review. There are many disagreements concerning the validity of these instruments and scales, and whether the patient information is reliable for the prescription of an adequate dose of analgesic medication. This situation was taken so seriously that the publication of an article considering "pain as the fifth vital sign" was withdrawn (Retraction. Regarding: Walid MS et al. 2008).

In the United States, this happened due to the conjunction of three factors: (a) the widespread diffusion of opioid medication for pain relief occurred at the same time; (b) an intense marketing program, with the opening of pain groups and clinics and false information about opioid analgesics; and (c) the growing greed of the pharmaceutical industry to make more and more profits. The potentiation between the three gave rise to a serious opioid epidemic, which goes from 2000 to the present moment, causing the addiction of many patients, as well as many deaths from overdose (Keefe 2022).

Although the disclosure about these facts took a long time to occur, so as not to contradict interests linked to profits, the evaluation and management of pain is still open [see Dydyk and Grandhe (2023) for an updated review on "pain assessment"].

The Central Nervous System

From a neuroanatomical and functional point of view, the NS comprises two large divisions: the *central nervous system* (CNS), and the *peripheral nervous system* (PNS) (Brodal 2010).

The CNS is formed by structures nested in the *encephalon, the brain,* and the *spinal cord*. From a functional point of view, the *brain*, housed in the *encephalon*, integrates most of the sensory functions, and coordinates mental, bodily, and behavioral functions. The spinal cord conducts signals between the brain and the body (Brodal 2010).

Together, the CNS and the PNS are responsible for both the adaptation of the organism to the external world and the regulation of its internal environment and *viscera*. For this task, they search and receive information from the sensory organs, and control muscle activity. In the human case, previous experiences and cognitive information play a relevant role in the evaluation of the lived situation (Brodal 2010).

When the CNS considers that a stimulus is relevant to preserve life, or for the quality of life of the human being, it will be stored in its long-term memory (Lynch 2004) for eventual later use. The neuroscientist Joseph LeDoux, who has studied the neural system of fear in detail, notes that threatening situations that cause fear, and are fundamental for survival, tend to be recorded in memory, even if they have occurred only once (LeDoux 1996).

Stimuli are selected by the CNS when they present some kind of *salience* that makes them stand out among others (Melloni et al. 2012; Chen et al. 2016) to the human being, or to some organ of it, for being threatening, urgent, or necessary, or still, very pleasant, desirable, or pleasurable to the being (De Houwer and Hermans 2001; Panksepp 1998).

Stephen W. Porges, in his *Polyvagal Theory*, considers that the CNS continuously evaluates the *risks* through the processing of sensory information that comes from the environment and the viscera (Robinson and Gebhart 2008; McNaughton and Corr 2018), especially those of a *social nature* (Porges 2009). Risk assessment operates like a radar in an observation field, spinning and trying to capture signals: at each change in the environment, the radar sends a signal indicating that the organism needs to readjust itself, which it does quickly by changing its routes and neural connections, thanks to *neural plasticity* (Radley et al. 2011). This risk assessment process is dynamic and continuous; we are adapting all the time.

As this type of neural assessment does not require *conscious perception* and can make use of information that comes from the *subcortical structures*, the author introduced the term *neuroception*—different from *perception*—which is the ability to unconsciously detect, in the environment, aspects with survival value (Porges 2011, p. 5).

The Human Brain

The *brain* is the master organ of the nervous system. It is made up of about 86 billion neurons, according to a recent estimate made by Suzana Herculano-Houzel and Roberto Lent, two prominent Brazilian neuroscientists; this amount is less than what they initially estimated (Herculano-Houzel and Lent 2005).

The human brain is a highly complex organ, the most evolved in the animal kingdom. Although it maintains more primitive aspects very similar to other mammals in subcortical structures, in higher order (cortical) structures, the brain is capable of unprecedented and sophisticated constructions (Bassett and Gazzaniga 2011; Hofman 2014).

In its evolution, the brain was one of the organs of the human body that underwent the greatest changes, quantitatively as well as qualitatively. The brain of the contemporary human being (Homo *sapiens sapiens*) is the result of progressive changes that occurred over about 50 million years since the appearance of the species *Homo* (Ribas 2006).

During this period, the total size of the brain had a significant increase in volume in relation to the size of the human body, a phenomenon called *encephalization* (Narayan and Verma 2019). Its volume more than tripled since the human being separated from the species of chimpanzees, about 7.1 to 7.2 million years ago; there was also a remarkable increase in cortical regions in relation to subcortical ones, and, within the former, those of the neocortex and the prefrontal cortex (PFC) (Stout and Chaminade 2012). The brain expansion occurred heterogeneously and was accompanied by changes in its structural complexity (Vallender et al. 2008).

Although there seems to be no doubt that the human brain is the most evolved organ in nature—due to its cognitive, language, abstraction, event anticipation, planning capabilities, etc.—, it is still not clear which factors contributed to its evolution and exceptionality.

One of the difficulties in studying brain evolution is that we cannot directly study ancestral brains, as their soft parts did not survive over time. However, some important information has been obtained in paleoneurological studies, made with endocranial molds; these are obtained by filling with resin the hollow places corresponding to the internal surface of the cranial box. These molds allow the reproduction of external morphological aspects of the brain, which enable the comparison of hominid brains with those of monkeys, measuring and evaluating the modifications in the areas of the frontal lobes. In hominid and human brains, a reorganization of the frontal lobes can be inferred throughout evolution, linked to cognitive functions, language, abstract thinking, planning of future actions, execution of motor activities, and decision-making (Falk et al. 2000).

Until recently, it was presumed that the privileged capabilities of the human brain were due to encephalization and the increase in cortical areas, as well as the mastery of fire and the consequent modification of the diet (Burini and Leonard 2018; Belo et al. 2017). Another hypothesis proposed that there could be evolutionary links between language, gestures, and tool use (Bradshaw and Nettleton 1982). None of these hypotheses were conclusive, but the development of recent study methods brought some new ideas, sometimes questioning the previous ones.

Another way to circumvent the impossibility of direct access to ancestral brains has been the study and comparison between the contemporary human brain with brains of less evolved species, like rodents and primates. The mouse brain has been used comparatively by many scientists and, in the words of the well-known Yugoslav neuroscientist based in the United States, Pasko Rakic, whose research focus is the evolution of the human brain, and this choice is due to it being "the best and most economical experimental model system" (Rakic 2009).

An alternative approach consists of directly studying the brain of contemporary man using postmortem donated brain tissues and stored in brain banks that have existed since the 1950s (Carlos et al. 2019).

The recent development of functional neuroimaging, and molecular biology, has opened a promising research front, which can be associated to other methods.

Currently living in the United States, the Brazilian neuroscientist Suzana Herculano-Houzel, who works with neuron counting research, argues that neither encephalization nor the size of the brain can account for its cognitive evolution, as its development did not diverge from that of other primates in terms of linear scale; in these terms, it would be just a more developed primate brain. From her research, Herculano-Houzel (2009) proposed that the key element in cognitive evolution would be the absolute number of neurons existing in the human brain.

In the same line of reasoning, Gabi et al. (2016), in a study in 2016, stated that there was no relative expansion in the number of prefrontal neurons in primate or human evolution, in the distribution of cortical neurons in primate or human evolution, nor visible differences between primates and humans in the distribution of cortical or glial cells along the midline of the head, and that, therefore, the most distinctive aspect of the PFC would be its absolute number of neurons. According to these authors, the bigger the absolute number of neurons, the more combinatorial interactions would be possible at synapses. This ability could grow exponentially, enabling more elaborate mental constructions, such as social cognition. Besides the number of neurons, the authors add genetic factors and life experience (epigenetic) as enrichers of brain skills.

The Brazilian neuroscientist Alianda Cornélio and others investigated another theory based on a mathematical model, which credited the development of the hominid brain to the mastery of fire and the consequent alteration of the diet. After examining the evidence, they proposed that evolution was probably due to the efficient gathering activity of our ancestors, more than feeding (Cornélio et al. 2016). The brain would be a case of enrichment by use. New methods in genomics and molecular biology have opened new and fascinating investigative possibilities to support evolution, among which the hypotheses of changes in the structure of the *genome*, *cell proliferation* in certain brain areas, changes in *cell behavior* (e.g., appearance and disappearance of genes) (Kristan Jr. 2016), alterations of *signaling pathways* and patterns of *gene expression*,[1] adaptive evolution of *proteins*, *neuronal migration* to certain brain regions, and differentiated *phenotypic expression* (Nachtomy et al. 2007; Lui et al. 2011).

Throughout its evolution, the brain was one of the organs of the human body that underwent the greatest changes. Qualitatively, the specialization of the prefrontal cortex translated into an enormous ability to deal with more complex and abstract concepts, create a culture, and develop a cultural accumulation, based on the collective effort of many generations, which is a typical human capacity (Stout and Chaminade 2012). The adaptability and flexibility of the prefrontal cortex functions was what enabled, during the 2020 pandemic, the quick adherence of the population to virtual language, online work, and social relationships through technology, even those notoriously averse to the digital world. For many, in fact, this was a matter of professional survival. From a technological point of view, the development of digital databases, with an increasingly larger data storage capacity, opened possibilities for the development of Artificial Intelligence.

Perhaps the most notable feature in the evolution of the brain is the fact that it took place in the direction of a *more sophisticated social behavior*, less violent, toward a *social regulation*, that is, contact and coexistence with others can help in the regulation of the person and society as well (O'Connell and Hofmann 2011).

To add to its new capabilities, changes occurred in the skull, in the musculature and in the anatomy of the face and neck, with greater control of the eyelids, aiming

---

[1] Gene expression is the generation of a phenotype from the hereditary information contained in a specific gene and some specific environmental stimuli. Phenotypes, in turn, are the observable characteristics resulting from the interactions of a genotype with the environment.

at their emotional expression; there was a migration of the eyes to the front of the face, facilitating eye-to-eye contact; and an improvement in the muscles of the middle ear allowed the human voice to be extracted from background sounds, contributing to a sophisticated social connection system (Porges 2011).

The Peripheral Nervous System

The peripheral nervous system (PNS) is formed by 12 pairs of *cranial nerves* and 31 pairs of *spinal nerves, ganglia* (clusters of neurons), and *nerve endings,* which are in the body outside the CNS, but penetrate the skull and spinal cord, respectively, and connect with them, receiving and sending messages between them. The spinal cord receives sensory information from the skin, joints, and muscles of the trunk and limbs, and responds with voluntary movements and reflexes. It also has ganglia that command the viscera (Akinrodoye and Lui 2023).

The PNS has two divisions: voluntary nervous system (VNS) and autonomic nervous system (ANS).

The VNS innervates the striated skeletal muscles, largely through the person's volition, performing the motor actions necessary for what the person wants to do.

Sympathetic Autonomic Nervous System (SANS), Parasympathetic (PANS), and Enteric Nervous System (ENS)

The ANS involuntarily innervates the *viscera,* the *smooth muscles,* the *exocrine glands,* and the *striated cardiac muscle.* Its relationships with emotions vary greatly, depending on the stimuli received and the individual's subjectivity (Kreibig 2010).

The ANS also regulates physiological processes (Ewing and Parvez 2008) and stress (Salim 2017) and is subdivided into *sympathetic nervous system* (SNS) and *parasympathetic nervous system* (PNS).

Several authors also consider a third subdivision of the ANS, the *enteric nervous system*, which controls the smooth muscles of the intestine (Nomaksteinsky et al. 2013).

The SNS manages the body's response to stress, coordinating the *general activation of metabolism* of various organs such as the increase in heart rate, the increase in glucose production, the dilation of the bronchi, and any other mechanisms necessary to produce energy and face challenges and dangers. In doing so, the SNS prepares the body for confrontation, although this implies a large expenditure of energy and immune depletion, to gather resources and protect what the body considers vital (Alshak and Das 2023; Chu et al. 2023).

In professional life, especially in the business world, with pressures to deliver work on tight deadlines and a large volume of tasks, professionals often automatically activate their stress mechanisms, as if that meant saving their own life. Consider the physical and mental wear and tear that the body must suffer under these circumstances.

The PNS, on the other hand, acts in the opposite direction, *protecting and restoring* the body's resources, deactivating, or inhibiting the mechanisms that the SNS had activated (Kandel et al. 2012). Most of the time, both systems have antagonistic functioning.

In addition to integrating all these specialties, the NS maintains reciprocal connections with the endocrine (glands and hormones), immune (body's defense system), and psychological (cognition, feelings, and behavior) systems, influencing them and being influenced by them, simultaneously (Manley et al. 2018).

Cortical and Subcortical Systems

In neuroscience, there is a consensus according to which the *phylogenetic* evolution of the *human brain,* which occurred over about six million years, took place from bottom to top—the superior structures were added later, and stacked on the lower ones; these latter were not eliminated, but had their action integrated with the superior, more recent ones (de Waal and Ferrari 2010).

The *cortex* (which in Latin means *bark*) is the outermost part of the brain, with a characteristic appearance full of folds, fissures, grooves, and convolutions. From the frontal part of the cortex, also called the *prefrontal cortex*, or *neocortex*, arise the most evolved and specifically human cognitive mental possibilities, such as the ability to *think before acting*, thanks to an inhibitory quality (which can be activated by conscious reflection, or by learning) (Konishi et al. 1999); the ability for *abstract thinking* (Dumontheil 2014); the ability to *think, remember, and speak* (Gabrieli et al. 1998); and the ability to *plan in time and space* (Tanji and Hoshi 2001).

These possibilities will not necessarily emerge in all human brains, being dependent on maturation conditioned by genetic, developmental, and cultural aspects. Some humans never reach this point.

The *subcortical system*, as the name indicates, is located below the cortex, and it is composed of various structures with specific functions, which relate to each other and with the rest of the brain. The subcortical structures are present in the brains of all mammals, including humans. An essential characteristic of them is that they produce the *primary affective-emotional contents,* which occur *unconsciously* (Panksepp and Biven 2012).

The subcortical structures, although not identical in the different mammalian species, maintain a remarkable homology among themselves (they are similar and have organ equivalence). This happens because they have been well preserved during evolution (Brodal 2010), which means that they were and continue to be efficient in the survival of mammals over many millennia—otherwise, they would have been modified, eliminated, or the species would have been extinct.

The homology between the subcortical areas of the brains of lower mammals has allowed comparative studies that help to clarify and anticipate many aspects of human brain affective neuroscience (Brodal 2010).

Many people intuitively, but erroneously, believe that the superior structures are always in command of the inferior ones, with the dominance of reason and consciousness over emotions and instincts. This notion, quite predominant in Western culture, is not supported neuroscientifically, and has given rise to many misconceptions, including in the sciences themselves.

On the other hand, it has been demonstrated through studies of functional magnetic resonance imaging (fMRI) that many decisions are made *before* the person is aware of them, and that these do not always depend on their volition or their conscious perception of the decision-making process (Smith 2008). Currently, computational models of decision-making system simulation have already been used as technological resource in the study of this neuroscientific area (Kirsch 2019).

Some people may habitually function in the predominance of the cortical command (more rational, thought out, and planned), in a type of processing called *top-down* or *descending* (concerning the direction of messages from the cortical to the subcortical area). On the other hand, some people may function in a more primitive way, from the subcortical system (emotional, instinctive, impulsive, unconscious), in the so-called *bottom-up* or *ascending* processing (Rauss and Pourtois 2013).

The construction of reality by the human brain is generally the product of the dynamics between the processing of ascending and descending messages in the brain, which integrates genetic factors, life history, and cultural aspects; primitive and more mature aspects; and more conscious or less conscious aspects (Sherman et al. 1997).

In moments of stress, threat, or danger, even the most mature individuals may momentarily lose their rational capacity, and act from the functioning of the subcortical structures—more impulsive, emotional, and unconscious, as mentioned, but faster in reaction, due to their antiquity and phylogenetic wisdom, with a greater chance of ensuring survival. Interestingly, the survival response may not be a socially appropriate response, as the pressure of the threatening situation tends to favor the immediate action of the survival and risk assessment circuits involved (Panksepp and Biven 2012). In this case, the response may be more crude, coarse, poorly elaborated, and only solve the immediate problem.

The survival mechanisms in human beings facing dangerous situations, consist in automatically triggering the behaviors of fight-or-flight, fainting, or immobilization [*freezing*], in addition to the physiological mechanisms related to energy storage and nutritional inuts (Porges 2011).

For instance, an employee who plans to ask for a raise at work may function in a much more cautious, programmed, and slow manner, and use more of his conscious cortical system to plan actions and achieve his goal (cortical predominance). But if this same person is robbed by a thief, he may tend to react in a much faster and automatic way, as he is about to defend his own life (subcortical response). Paradoxically, this more immediate response may have the opposite effect to the desired one—a sudden movement, a shot, and life is lost.

The instinctive subcortical impulses, in addition to being unconscious, can possess a very intense force. Under these circumstances, a person can be led to act in a way that contradicts their rational or moral convictions, without even understanding or explaining their actions. The movie *Wolf* (1994), starring *Jack Nicholson,* poetically and emblematically, although terrifyingly, illustrates the force of instincts over the "civilized" aspects of the human being—the transformation of Nicholson into a werewolf.

There are in our language various semantic expressions used as metaphors, but which, in fact, translate the real archaic/primitive content of some situations: *"I lost my mind!" "I was blinded by rage," "I wanted to kill so-and-so,"* etc. Instead of saying in vino veritas, we would better say *in metaphora veritas*.

Still considering subcortical functioning, it is common to find, in clinical consultation, certain dynamics of action and primitive mental functioning that are part of the patient's family culture, whose members unconsciously share it among themselves, without realizing it. A patient with higher education reported a great fear of dead people, the possibility of their return, supernatural appearances, and ideas of the sort, exacerbated by the pandemic. I initially thought about the hypothesis of a psychotic symptom, but, when we investigated her life history, we found that she was born and raised in a very small town in the countryside, where the inhabitants shared folklore on this subject in a very similar way to her discourse. In this case, her higher education was not being effective in filtering out the primitive and delusional family ghosts, and, at that moment, she could not operate within the principle of reality.

In neuroscience, there is still no consensus on how human cortical and subcortical structures operate, and how they interact with each other, with several hypotheses being proposed.

From his many animal studies, Jaak Panksepp proposes that the brain-mind is hierarchically structured in three evolutionary levels, from bottom to top. The first level, the oldest structure in phylogenetic terms, is situated at the bottom of the neuroanatomical brain architecture, which he names as the *system of primary processes*, would be responsible for instinctive emotional responses; these would be raw affects and emotions, without cortical representation, nameless, outside of symbolic thought. The *system of secondary processes* would be mainly formed by learning and memories, object relations and intersubjectivity, and is situated at the level of the basal ganglia; apparently, they are "deeply unconscious" (Panksepp and Biven 2012, p. 9–10). Finally, the *system of tertiary processes* would produce the highest order cortical processes, the most sophisticated of the human mind: abstract constructions, symbolic thought, self-awareness, for example.

Neuroscientist Daniel S. Levine (2017), in a recent article, considers that, in the last 50 years, mainly after Paul MacLean's theory of the *triune brain* (MacLean 1988), brain researchers have agreed on a tripartition of the mind into *instincts, emotions,* and *thoughts*, with extensive areas of overlap and connections between the brain regions involved in the three. He also observes that emotional processes are not always automatic, and, in general, are not opposed to reason. Levine uses neuroanatomical brain structures involved in decision-making as a reference to substantiate his arguments and concludes that an appropriate decision requires all three elements.

Although basic emotions and affects may be unconscious to a person, it can be more easily perceived by an observer (another person, or even a pet), from some change in facial or bodily expression, verbalization, or nonverbal signs.

An interesting example occurred in our doctoral research. In the final interview, a volunteer who had been administered OT narrated that one day, during the research, he was in the car with his wife and started singing. She then observed: *"Wow, you're different! You're singing! But you never sing!"* His wife noticed the change before him (Nogueira-Vale 2019).

A final, but decisive, observation about the brain location of the emergence of emotions is made by Panksepp: throughout his research, he observed that, for the most part, there were no emotional manifestations when he stimulated the animal's cortical area with an electric current. In the subcortical area, however, the electrical stimulation of a given circuit clearly elicited an emotion, and always the same one (Panksepp and Biven 2012, p. 26).

Left and Right Hemispheres

Another way to divide the brain is by a medial sagittal cut, in which a *left hemisphere* and a *right hemisphere* can be observed, connected by a brain structure called the *corpus callosum*. The two hemispheres are not identical, and some functional differences between them will be later addressed.

Neurons and Glial Cells

The *neuron*, or nerve cell, is the basic neuroanatomical unit of the nervous system (NS). However, it is not its functional unit, although this is often stated, since an isolated neuron has no function at all: the functional unit would be a *group, or an assembly of neurons*.

In addition to neurons, there is another class of nerve cell in the brain called *glia*, or *glial cell,* which we will briefly discuss. Until recently, glial cells were considered as a simple support for nervous tissue. Research conducted in the last decade, however, discovered that they have several important features: they can function as *stem cells* and *progenitor cells,* capable of generating different cells from themselves, maintaining the vascular tone, and modulating the synaptic environment; and help in neural migration. They can also have an immunological function in cases of neurodegenerative diseases, infections, or injuries. Other glial cells are responsible for the myelination of neuronal axons (Hughes 2021).

In neuroanatomical terms, and in a simplified way, each neuron is formed by a *membrane,* a *body* with a *nucleus, dendrites* in the form of branches, and an *axon,* which is a single bundle thicker than the dendrites, with a nerve branch at its end.

To perform the primary function of sending *neural messages* throughout the body, a neuron needs others with which to communicate and associate for more complex actions. Groups of neurons can be associated with each other, forming more permanent basic circuits, and can also form temporary sets (assemblies) with other neurons (Gerstein et al. 1989), thanks to *neural plasticity,* in response to specific needs (von Bernhardi et al. 2017).

Neural plasticity allows not only a neural rearrangement in case of injury, but also a constant alteration of neural networks, to meet the dynamic needs of life (Zhong et al. 2023).

All this "conversation" between neurons is extremely fast, a zigzag of conscious and unconscious messages, *top down* and *bottom up*, which can either compete with each other or add up for a convergent outcome (Mysore and Kothari 2020).

Imagine a young man who argued with his girlfriend and is anxious to talk to her, but afraid to call or send a message. He rehearses, tries to call, and then hangs up the phone. He may eventually call. He may give up on calling. Or he may throw his cell phone against the wall. Throughout this situation, his temperature may rise, he may sweat, have accelerated heartbeats, get a dry mouth, have jumbled and confused ideas, secrete cortisol and catecholamines, contract his muscles, get angry, scared, anxious, just to mention some of the possible *psychoneuroendocrinological* interactions that can occur simultaneously. Can you imagine all the excitatory and inhibitory messages that must have occurred in his brain, and, consequently, in his body and behavior in this process?

Neural Messages

The *neural messages* command the functioning of the body and the psychism of human beings, indicating what these should do at a certain moment to meet the demands of life. They are generated from stimuli that come from the internal environment (the being itself), and external (environment, other people), and are sent in the form of *nervous impulses* through a *neural network* to their final target.

Neural messages are conducted by *electrical stimuli* (which occur within the neuron) and *chemicals* (between neurons). The final target of the message could be another neuron, a muscle cell, a gland, or an organ. The pulsation of neural messages keeps the body functioning all the time, even when we are sleeping, being responsible for the maintenance of important autonomous functions (Vyazovskiy et al. 2009).

When any event disturbs the organism, and as a result, the resting state of a neuron is broken, an *action potential*, or electrical pulse, is created, responsible for the eventual firing and propagation of a neural message through the body, which sometimes must travel long distances to reach its final target (Krueger-Beck et al. 2011).

The neural message enters the neuron through the membrane, crosses the body cell carried by electrical signals, and follows the axon until it reaches a point of approach with the next neuron.

Since neurons do not touch each other, being separated by a narrow space called *synaptic cleft*, it is necessary that the message, after crossing the so-called *presynaptic* neuron, moves through the cleft to reach the next neuron, called *postsynaptic* (Kandel et al. 2012).

Within the cleft, there is an active zone for transmitting nerve impulses from one neuron to another, called *synapse*. At the synapse, the electrical transmission will be exchanged for chemical transmission, and different substances (neurotransmitters, neuromodulators, neuropeptides) will be released through existing channels in the membrane of the presynaptic neuron. These substances will drive the messages across the cleft, until they penetrate the *receptors* that exist in the membrane of the postsynaptic neuron (Kandel et al. 2012).

There is a prerequisite, however, for the entrance of a neurotransmitter substance carrying a message into the receptor of the postsynaptic cell: it can only do so if the cell has, on its membrane, *specific receptors* for that substance, a design analogous to the fitting of a key (substance) into a lock (receptor) (Lovinger 2008).

Synapses can be *excitatory,* when they prepare the organism for an action—if I feel hungry, I receive messages from the body-mind to go after food. When the synapse is *inhibitory*, on the contrary, it will inhibit the action or functioning of the person or an organ, such as when a person sees a scorpion and withdraws his/her hand (Robinson and Gebhart 2008).

This is also what happens when the heart exhibits an excess of heartbeats: the vagus nerve, inhibitory, is activated, sending a message for the heart to slow down (inhibit) the heart rate, until it returns to normal rhythm, performing a cardiac regulation. In this type of regulation (autonomous), the action is totally automatic, and does not depend on initiative or any conscious act of the person (Li et al. 2004).

If the stimuli received by the body-mind have sufficient action potential to trigger the neural message, the outcome can vary from reflex actions to thoughts, reasoning, affective reactions, motor actions, and other complex behaviors.

When the nerve impulse is below the action threshold, it will not advance to the next neuron, but it may still *sensitize* it. As a result, if a new stimulus occurs, the message may be triggered even if it is below the action potential, a sign that it was stored in the organism's memory, ready for action a posteriori (Robinson 2010). This storage will occur if the organism understands that this information is vital.

Imagine an allergic person exposed to an allergenic substance, pollen, for example: the first time there may not be a visible reaction in the body, but after new exposures, the nose will start to itch, sneeze, and run, until it unfolds into a full allergic rhinitis, due to an immune response (Andersson and Tracey 2012).

Neurotransmitter Substances

The neurotransmitter substances, whose role is to diffuse the neural messages in the synaptic cleft and send them to the next neuron (postsynaptic), are generically classified as *neurotransmitters* (they transmit the message), *neuromodulators* (increase or decrease the intensity of the message), or *neurohormones* (the message is emitted by glandular cells and enters the bloodstream, sometimes traveling long distances in search of the target). There are disagreements among authors regarding this classification of neurotransmitter substances, which do not have very precise definitions or frontiers (Panawala 2017).

Neuropeptides

Among the neurotransmitter substances, we will highlight, for the purposes of this book, the important class of *neuropeptides*.

The discovery of the first *neuropeptides*—small molecules produced and released by neurons in the synaptic cleft—took place in the early 1950s and contributed decisively to a revolution in the understanding of the human being (Klavdieva 1996). However, the term *neuropeptide* was only mentioned for the first time in the early 1970s by David de Wied, when knowledge about them was still in its infancy. Since then, their study has evolved greatly, with the support of computational technologies and databases (Burbach 2010).

In the 1970s, the study of neuropeptides took a leap, thanks to the development of new analysis techniques, with emphasis on the *radioimmunoassay*, developed by American medical physicist Rosalyn Yallow, who, for this work, received the Nobel Prize in 1977 (Nobel Prize, org.). Thanks to these discoveries, there was a real explosion of experiments with animals, and a significant database began to form. Initially only half a dozen neuropeptides were identified. By 1977, it was known that they are encoded by more than 70 genes in the genomes of mammals (Sherman et al. 1997).

And why did neuropeptides cause such a revolution? When it was discovered that they participated in various brain functions related to the control and modulation of behavior, especially *motivational* and *emotional processes*. In addition, neuropeptides control and coordinate the *autonomous* (i.e., automatic) functioning of *visceral organs* (Cervero 2000), whose neural messages can be transmitted to the cerebral cortex (where they may be symbolized, verbalized, and enter consciousness), influencing the way humans process and react to new information. Thus, complex mind-brain-body patterns are created, which the renowned American neuroscientist Candace Pert (1946–2013) appropriately called *psychosomatic network* (Pert et al. 1985).

This discovery was fundamental for the progress in the study of the articulations between body, mind, and affects, opening the field for an affective neuroscience with new scientific bases (Pert et al. 1985; Panksepp 1998; De Houwer and Hermans 2001) after more than half a century in which affects were ignored or undervalued by neuroscience.

Unfortunately, to this day, despite scientific evidence, certain cognitive researchers continue to primarily investigate cognitive-behavioral aspects and ignore affective-emotional aspects, which contribute to disseminating an incomplete or imprecise view of the human being. An important aspect of psychic phenomena is then disregarded because, literally, they have no name: manifestations that announce themselves by blushing, by paling, by discomfort, by the weakening of the knees, by manifestations on the skin, etc. Such are some manifestations of primary emotions.

Neural Systems

Now that you have learned how the neuron and neural messages work, we can move on to a slightly more complex structure, called *neural system.*

Neural systems are sets or networks of specialized neural circuits, which evolved in the human body to generate certain specific responses.

Among the neural systems, we highlight the class of *affective neural systems*, which provoke emotional reactions, affective states, and specific moods in the person.

Recalling what we said, the affective state generated can be *positive* (hedonic) or *negative* (unpleasant, anguishing, frightening, enraging, for example). This response is invariant for each network, or from an association between networks, although it can be modulated to different intensities. It is invariant because whenever this neural system is activated, it will give the same type of affective response. In addition to being activated by stimuli from the environment, neural systems can also be activated by an electric current, via an electrode inserted at the site of the specific neural system, or by infusion of neuroactive substances into the brain. Here we are talking about animal experiments in the 1970s, emphasizing that these procedures would not be ethically acceptable in most experiments with humans (Panksepp 1998).

For Panksepp and Biven (2012), these affective neural systems are activated independently of the person's will, and function in an autonomous and unconscious way.

Each affective neural system, composed of a neural network, is activated and modulated by its respective neuropeptide(s).

The unraveling of the neuroanatomy and functioning of the basic affective neural systems was only possible thanks to the development of more powerful neuroimaging, new methods in clinical trials, and many detailed works of investigation and dissection in animal brains (Panksepp 2015).

Panksepp's Basic Affective Neural Systems

Jaak Panksepp, in his theory and experiments on affective neuroscience (1998), identified seven basic affective neural systems. Some of them, more archaic from an evolutionary point of view, and present in some lower animals, are SEEKING System (territorial exploration), LUST (search for partners sexual and for procreation purposes), RAGE (competition and defense of resources), and FEAR (attack-and-flight to protect physical integrity). The most evolved affective neural systems are linked to social skills: CARE System (with offspring), PANIC (linked to situations of abandonment and helplessness), and PLAY (social activities for bond formation).

By introducing this material on the nervous system, we hope to have conveyed to the reader an idea of the extent and dimension of nervous phenomena, the enormous number of neural mechanisms, synapses, and messages that travel through circuits and neural systems simultaneously, creating a huge three-dimensional network in action.

References

Akinrodoye MA, Lui F (2023) Neuroanatomy, somatic nervous system. In: StatPearls [Internet]. StatPearls Publishing, Treasure Island. Available from: https://www.ncbi.nlm.nih.gov/books/NBK556027/

Alshak MN, Das J (2023) Neuroanatomy, sympathetic nervous system. In: StatPearls [Internet]. StatPearls Publishing, Treasure Island. Available from: https://www.ncbi.nlm.nih.gov/books/NBK542195/

Andersson U, Tracey KJ (2012) Reflex principles of immunological homeostasis. Annu Rev Immunol 30:313–335

Bassett DS, Gazzaniga MS (2011) Understanding complexity in the human brain. Trends Cogn Sci 15(5):200–209

Belo LLA, Teles KI, Silva HM (2017) Efeitos da alimentação na evolução humana: uma revisão. Conex Ciência (online) 12(3):93–105

Bradshaw JL, Nettleton NC (1982) Language lateralization to the dominant hemisphere: tool use, gesture, and language in hominid evolution. Curr Psychol Rev 2:171–192

Brodal P (2010) The central nervous system: structure and function, 4th edn. Oxford University Press

Burbach JP (2010) Neuropeptides from concept to online database www.neuropeptides.nl. Eur J Pharmacol 626(1):27–48

Burini RC, Leonard WR (2018) The evolutionary roles of nutrition selection and dietary quality in the human brain size and encephalization. Forum Nutr 43(19)

Carlos AF, Poloni TE, Medici V, Chikhladze M, Guaita A, Ceroni M (2019) From brain collections to modern brain banks: a historical perspective. Alzheimers Dement (N Y) 5:52–60

Cervero F (2000) Visceral pain-central sensitization. Gut 47(Suppl 4):iv56–7. discussion iv58

Chen C, Zhang X, Wang Y, Zhou T, Fang F (2016) Neural activities in V1 create the bottom-up saliency map of natural scenes. Exp Brain Res 234(6):1769–1780

Chu B, Marwaha K, Sanvictores T et al (2023) Physiology, stress reaction. In: StatPearls [Internet]. StatPearls Publishing, Treasure Island. Available from: https://www.ncbi.nlm.nih.gov/books/NBK541120/

Cornélio AM, de Bittencourt-Navarrete RE, de Bittencourt BR, Queiroz CM, Costa MR (2016) Human brain expansion during evolution is independent of fire control and cooking. Front Neurosci 10:167

de Houwer J, Hermans D (2001) Automatic affective processing. Cognit Emot 15(2):113–114

De Houwer J, Thomas S, Baeyens F (2001) Associative learning of likes and dislikes: a review of 25 years of research on human evaluative conditioning. Psychol Bull 127(6):853–869

de Waal FB, Ferrari PF (2010) Towards a bottom-up perspective on animal and human cognition. Trends Cogn Sci 14(5):201–207

Dumontheil I (2014) Development of abstract thinking during childhood and adolescence: the role of rostrolateral prefrontal cortex. Dev Cogn Neurosci 10:57–76

Dydyk AM, Grandhe S (2023) Pain assessment. In: StatPearls [Internet]. StatPearls Publishing, Treasure Island. Available from: https://www.ncbi.nlm.nih.gov/books/NBK556098/

Ewing GW, Parvez SH (2008) Neuro-regulation of the physiological systems by the autonomic nervous system – their relationship to insulin resistance and metabolic syndrome. Biogenic Amines 22(4–5):208–239

Falk D, Redmond JC Jr, Guyer J, Conroy C, Recheis W, Weber GW, Seidler H (2000) Early hominid brain evolution: a new look at old endocasts. J Hum Evol 38(5):695–717

Gabi M, Neves K, Masseron C, Ribeiro PF, Ventura-Antunes L, Torres L, Mota B, Kaas JH, Herculano-Houzel S (2016) No relative expansion of the number of prefrontal neurons in primate and human evolution. Proc Natl Acad Sci USA 113(34):9617–9622

Gabrieli JD, Poldrack RA, Desmond JE (1998) The role of left prefrontal cortex in language and memory. Proc Natl Acad Sci USA 95(3):906–913

Gerstein GL, Bedenbaugh P, Aertsen MH (1989) Neuronal assemblies. IEEE Trans Biomed Eng 36(1):4–14

Herculano-Houzel S (2009) The human brain in numbers: a linearly scaled-up primate brain. Front Hum Neurosci 3:31

Herculano-Houzel S, Lent R (2005) Isotropic fractionator: a simple, rapid method for the quantification of total cell and neuron numbers in the brain. J Neurosci 25(10):2518–2521

Hofman MA (2014) Evolution of the human brain: when bigger is better. Front Neuroanat 8:15

Hughes AN (2021) Glial cells promote myelin formation and elimination. Front Cell Dev Biol 9:661486

Kandel E, Koester JD, Mack SH, Siegelbaum SA (2012) Principles of neural science, 5th edn. McGraw Hill, Nova York

Keefe PR (2022) Empire of pain. Anchor Books

Kirsch A (2019) A unifying computational model of decision making. Cogn Process 20(2):243–259

Klavdieva MM (1996) The history of neuropeptides II. Front Neuroendocrinol 17:126–153

Klimek L, Bergmann KC, Biedermann T, Bousquet J, Hellings P, Jung K, Merk H, Olze H, Schlenter W, Stock P, Ring J, Wagenmann M, Wehrmann W, Mösges R, Pfaar O (2017) Visual analogue scales (VAS): measuring instruments for the documentation of symptoms and therapy monitoring in cases of allergic rhinitis in everyday health care: position paper of the German Society of Allergology (AeDA) and the German Society of Allergy and Clinical Immunology (DGAKI), ENT section, in collaboration with the Working Group on Clinical Immunology, Allergology and Environmental Medicine of the German Society of Otorhinolaryngology, Head and Neck Surgery (DGHNOKHC). Allergo J Int 26(1):16–24

Konishi S, Nakajima K, Uchida I, Kikyo H, Kameyama M, Miyashita Y (1999) Common inhibitory mechanism in human inferior prefrontal cortex revealed by event-related functional MRI. Brain 122(Pt 5):981–991

Kreibig SD (2010) Autonomic nervous system activity in emotion: a review. Biol Psychol 84(3):394–421

Kristan WB Jr (2016) Early evolution of neurons. Curr Biol 26(20):R949–R954

Krueger-Beck E, Scheeren EM, Nogueira-Neto GN, Button VLSN, Neves EB, Nohama P (2011) Potencial de ação: do estímulo à adaptação neural. Fisioter Mov 24(3):535–547. Licenciado sob uma Licença Creative Commons

References

Leal EM, de Serpa OD Jr (2013) Acesso à experiência em primeira pessoa na pesquisa em Saúde Mental [Access to first-person experience in research into mental health]. Cien Saude Colet 18(10):2939–2948. (Portuguese)

LeDoux JE (1996) The emotional brain: the mysterious underpinnings of emotional life. Simon & Schuster

Levine DS (2017) Modeling the instinctive-emotional-thoughtful mind. Elsevier

Li M, Zheng C, Sato T, Kawada T, Sugimachi M, Sunagawa K (2004) Vagal nerve stimulation markedly improves long-term survival after chronic heart failure in rats. Circulation 109(1):120–124

Lovinger DM (2008) Communication networks in the brain: neurons, receptors, neurotransmitters, and alcohol. Alcohol Res Health 31(3):196–214

Lui JH, Hansen DV, Kriegstein AR (2011) Development and evolution of the human neocortex. Cell 146(1):18–36

Lynch MA (2004) Long-term potentiation and memory. Physiol Rev 84(1):87–136

MacLean PD (1988) Triune brain. In: Comparative neuroscience and neurobiology. Readings from the encyclopedia of neuroscience. Birkhäuser, Boston

Manley K, Han W, Zelin G, David A, Lawrence DE (2018) Crosstalk between the immune, endocrine, and nervous systems in immunotoxicology. Curr Opin Toxicol 10:37–45

McNaughton N, Corr PJ (2018) Survival circuits and risk assessment. Curr Opin Behav Sci 24:14–20

Melloni L, van Leeuwen S, Alink A, Müller NG (2012) Interaction between bottom-up saliency and top-down control: how saliency maps are created in the human brain. Cereb Cortex 22(12):2943–2952

Mobbs D, Hagan CC, Dalgleish T, Silston B, Prévost C (2015) The ecology of human fear: survival optimization and the nervous system. Front Neurosci 9:55

Mysore SP, Kothari NB (2020) Mechanisms of competitive selection: a canonical neural circuit framework. Elife 9:e51473

Nachtomy O, Shavit A, Yakhini Z (2007) Gene expression and the concept of the phenotype. Stud Hist Phil Biol Biomed Sci 38(1):238–254

Narayan RK, Verma M (2019) Encephalization. In: Vonk J, Shackelford T (eds) Encyclopedia of animal cognition and behavior. Springer, Cham

Nogueira-Vale EA (2019) Relações entre ocitocina, apego e sono em pessoas com Transtorno de Ansiedade Generalizada [tese]. Instituto de Psicologia, São Paulo. citado 2024-01-17

Nomaksteinsky M, Kassabov S, Chettouh Z, Stoeklé HC, Bonnaud L, Fortin G, Kandel ER, Brunet JF (2013) Ancient origin of somatic and visceral neurons. BMC Biol 11:53

O'Connell LA, Hofmann HA (2011) The vertebrate mesolimbic reward system and social behavior network: a comparative synthesis. J Comp Neurol 519(18):3599–3639

Panawala L (2017) Difference between neuropeptides and neurotransmitters [internet]. Disponível em: https://www.researchgate.net/publication/318305704_Difference_Between_Neuropeptides_and_Neurotransmitters

Panksepp J (1998) Affective neuroscience: the foundations of human and animal emotions. Oxford University Press

Panksepp J (2015) Comunicação pessoal em 25 de agosto de

Panksepp J, Biven L (2012) The archaeology of mind: neuroevolutionary origins of human emotion. Norton series in evolutionary biology. W. W. Norton & Company

Pert CB, Ruff MR, Weber RJ, Herkenham M (1985) Neuropeptides, and their receptors: a psychosomatic network. J Immunol 135(2 Suppl):820s–826s

Porges SW (2009) Reciprocal influences between body and brain in the perception and expression of affect: a polyvagal perspective. In: Fosha D, Siegel DJ, Solomon MF (eds) The healing power of emotion: affective neuroscience, development & clinical practice. W. W. Norton & Company, Nova York, pp 27–54

Porges SW (2011) The polyvagal theory: neurophysiological foundations of emotions, attachment, communication, and self-regulation. Norton Series on Inte. Personal Neurobiology

Radley JJ, Kabbaj M, Jacobson L, Heydendael W, Yehuda R, Herman JP (2011) Stress risk factors and stress-related pathology: neuroplasticity, epigenetics and endophenotypes. Stress 14(5):481–497
Rakic P (2009) Evolution of the neocortex: a perspective from developmental biology. Nat Rev Neurosci 10(10):724–735
Rauss K, Pourtois G (2013) What is bottom-up and what is top-down in predictive coding? Front Psychol 4:276
Retraction. Regarding: Walid MS, Donahue SN, Darmohray DM, Hyer LA Jr, Robinson JS Jr (2008) The fifth vital sign-what does it mean? Pain Pract 6:417–422. Pain Pract. 2009;9(3):245
Ribas GC (2006) Considerações sobre a evolução filogenética do sistema nervoso, o comportamento e a emergência da consciência [Considerations about the nervous system phylogenetic evolution, behavior, and the emergence of consciousness]. Braz J Psychiatry 28(4):326–338. (Portuguese)
Robinson TE (2010) Sensitization to drugs. In: Stolerman IP (ed) Encyclopedia of psychopharmacology. Springer, Berlin/Heidelberg
Robinson DR, Gebhart GF (2008) Inside information: the unique features of visceral sensation. Mol Interv 8(5):242–253. https://doi.org/10.1124/mi.8.5.9
Salim S (2017) Oxidative stress and the central nervous system. J Pharmacol Exp Ther 360(1):201–205
Schraube E (2014) First-person perspective. In: Theo T (ed) Encyclopedia of critical psychology. Springer, New York
Sherman SL, DeFries JC, Gottesman II, Loehlin JC, Meyer JM, Pelias MZ, Rice J, Waldman I (1997) Behavioral genetics '97: ASHG statement. Recent developments in human behavioral genetics: past accomplishments and future directions. Am J Hum Genet 60(6):1265–1275
Smith K (2008) Brain makes decisions before you even know it. Nature
Stout D, Chaminade T (2012) Stone tools, language, and the brain in human evolution. Philos Trans R Soc Lond Ser B Biol Sci 367(1585):75–87
Tanji J, Hoshi E (2001) Behavioral planning in the prefrontal cortex. Curr Opin Neurobiol 11(2):164–170
Vallender EJ, Mekel-Bobrov N, Lahn BT (2008) Genetic basis of human brain evolution. Trends Neurosci 31(12):637–644
von Bernhardi R, Bernhardi LE, Eugenín J (2017) What is neural plasticity? Adv Exp Med Biol 1015:1–15
Vyazovskiy VV, Olcese U, Lazimy YM, Faraguna U, Esser SK, Williams JC, Cirelli C, Tononi G (2009) Cortical firing, and sleep homeostasis. Neuron 63(6):865–878
Zhong Y, Zhou J, Li P, Jie G (2023) Dynamically evolving deep neural networks with continuous online learning. Inf Sci 646:119411

Chapter 4
Oxytocin

> *If one places a small, naturally occurring nine-amino-acid peptide called vasotocin into the brains of male frogs and lizards, they begin to exhibit courting sounds and sexual behaviors. Given the opportunity, males treated with vasotocin mount and clap females, and copulate. In mammals, two evolutionary offspring of these reptilian and piscine hormones, vasotocin and oxytocin, assume key roles in controlling certain aspects of sexual behaviors. Panksepp (1998, p. 230)*

OT and AVP: Molecular Structure, Neuroanatomical Location, Neurosecretory Cell

Oxytocin (OT) and its complementary hormone, arginine-vasopressin, also called vasopressin (AVP) or antidiuretic hormone, are small neuropeptides chemically composed of a *ring with six amino acids*: cysteine-tyrosine-isoleucine-glutamine-asparagine-cysteine, in addition to two *disulfide* (S) bonds between the two cysteines, *and a tail*, with three amino acids (proline, leucine, and glycine). AVP has a similar structure to that of OT, with only two different amino acids (Carter 1998). On the other hand, the two differ from *vasotocin,* which is an ancestral hormone, by a single amino acid each (Mahlmann et al. 1994). The gene housing the two occupies the same chromosome, with a functional interaction between them (Carter 1998); however, research done on OT is much more extensive than that on AVP, which still deserves to be more studied. It is estimated that the precursor protein of OT was encoded by an ancestral gene, about 500 million years ago (Gimpl and Fahrenholz 2001).

OT and AVP will be released throughout the *cell body* (also called *soma or perikaryon*), the *dendrites*, and the *collateral part of the axons* (Gimpl and Fahrenholz 2001). OT messages can be sent through the brain in two ways: (a) by *synaptic transmission*, along a neural pathway; (b) by *volume transmission* in nonsynaptic regions, spreading diffusely through the extracellular fluid between tissues, potentially reaching a much larger number of neurons in distinct locations in the brain (Fuxe et al. 2012).

Hypothalamus Hypophysis: Adenohypophysis and Neurohypophysis. Paraventricular Nucleus (PVN), and Supraoptic Nucleus (SON)

The *hypothalamus* corresponds to a small area located in the central region of the brain. Although it is approximately 4.3 cm^3 in size, it has fundamental regulatory functions in the human body (Hofman and Swaab 1992), and connections with all areas of the CNS, thanks to its privileged location. It is formed by several pairs of hormone-secreting nuclei (each pair is divided in one nucleus in the right brain hemisphere, and the other in the left hemisphere). One of its main functions is to maintain the homeostatic regulation of various physiological processes in the body, such as in *sleep and wakefulness, hunger and satiety, thermoregulation (regulation of body temperature), body fluids, networks of subcortical affective systems* (Esperidião-Antonio et al. 2017), *and the ANS*. Two types of these hypothalamic nuclei, the paraventricular nucleus (PVN) and the supraoptic nucleus (SON), are *secretory cells* in charge of the *production of neuropeptides OT and AVP*. In the PVN, both are produced by magnocellular cells; in the SON, production occurs from parvocellular cells (Fig. 4.1).

The *pituitary,* or *hypophysis,* is a brain region located just below the hypothalamus (Fig. 4.1) and has the shape of a bag with two sacks (or lobes). Although it appears to be a single structure divided into two halves, the lobes have distinct embryonic origins (Moore et al. 2020) and are anatomically and functionally different. The anterior lobe of the pituitary, also known as *adenohypophysis*, is an important gland responsible to produce *various trophic hormones* (i.e., precursors of other hormones), and *prolactin*, a hormone associated with the production of breast milk.

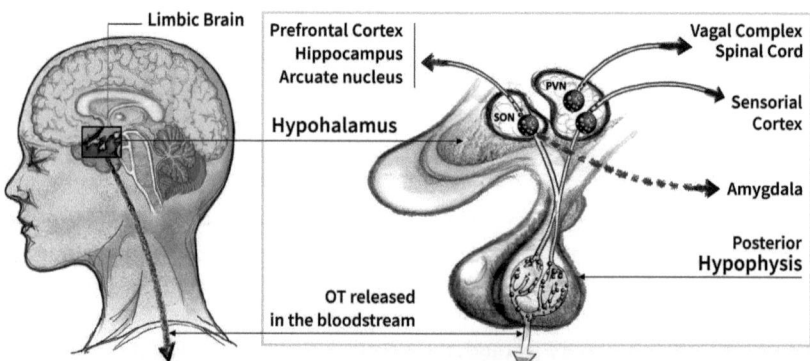

Fig. 4.1 OT is a neurohormone produced in magnocellular neurons located in the supraoptic nucleus (SON), and parvocellular neurons, located in the parvocellular nucleus. These neurons extend their axons through the infundibulum toward the posterior pituitary gland (neurohypophysis) where hormone is secreted and stored in vesicles. From there, the hormone is either released into the bloodstream, heading to target organs, or direct its fibers to central brain targets. (Illustration by Ciro Araujo)

In situations of stress, the adenohypophysis is responsible for initiating a cascade of hormones that end at the *adrenal glands* (or *suprarenal glands*), situated just above the kidneys, resulting in the production of the *stress hormones*.

The adrenal gland is triangular-shaped, with an outer part called *cortex*, and an inner part, the *medulla*. *Cortisol* is produced in the adrenal cortex and is one of the most important hormones for the response to stress. *Adrenaline (*or *epinephrine)* and *noradrenaline (*or *norepinephrine)* are produced in the adrenal medulla. These three hormones are responsible for organizing an integrated response to stress and constitute the *hypothalamic-pituitary-adrenal,* or *(HPA), axis* (Panksepp 1998).

The *neurohypophysis* (or *posterior pituitary*) is not a gland, as it does not produce hormones; it only stores them. It is a continuation of hypothalamic tissue and consists of neural fibers, hence called *pars nervosa*. The hypothalamic cells that secrete OT and AVP direct their axons to the *infundibulum*, which is a narrowing at the end of the hypothalamus and descends toward the posterior lobe of the pituitary. There, OT and AVP (produced in the PVN and SON nuclei of the hypothalamus) are transported by the axons and stored in secretory granules, which remain in the neurohypophysis until they receive a neural stimulus that releases them into the neurohypophyseal veins; from there, they go into the bloodstream (or systemic circulation). Through this route, they are distributed peripherally (i.e., outside the CNS) toward the target organs in the body, as opposed to the brain. The parvocellular OT cells also project toward the *pineal gland* and the *cerebellum*, continuing to the *spinal cord*, from where they will regulate autonomic functions (Gimpl and Fahrenholz 2001).

Distribution of Peripheral and Central OT

OT can be classified as peripheral, when its effects occur in the body, or central, when they occur in the brain.

The most well-known effects of peripheral OT are the contractile movements, in pulses, of the uterus and mammary glands, for the expulsion of the fetus at birth, and for milk ejection (Magon and Kalra 2011).

Peripheral OT is synthesized in tissues of the uterus, placenta, corpus luteum, amnion, testicles, and heart (Gimpl and Fahrenholz 2001), and of the kidney, thymus, pancreas, and adipocytes (Fontanez et al., retrieved on April 23, 2021). Given the quantity and quality of the tissues where it is produced, it is possible to have an idea of its plural effects.

Central OT is produced in smaller parvocellular cells, located in the PVN and other brain areas, and then released and distributed within the CNS itself (Uvnäs-Moberg 1996), toward various *brain target organs*:

- Olfactory bulb (very important in rodents, they have the functions of detecting predator odors, recognizing individuals of the same species, mating and sexual arousal).

- *Hippocampus* (memory).
- Other areas of the *hypothalamus*.
- *Raphe nucleus* (production of serotonin, which regulates various physical and mental functions).
- *Locus coeruleus* (physiological responses to stress and fear).
- *Substantia nigra* (dopaminergic cells linked to sleep and some motor behaviors).
- Cells of the vagal system (homeostasis).
- Spinal cord.
- Bed nucleus of the *stria terminalis* (BNST).
- Anterior commissure.

The BNST is, in fact, a continuation of the *centromedial amygdala*. It is an excitatory noradrenergic subcortical nucleus, with strong neural connections to the PVN, in the hypothalamus, where extensions of the stress axis, or HPA axis (Avery et al. 2016), and OT and AVP neurons are located. What would be the consequences of this neuroanatomical proximity between the neurons secreting stress hormones and the neurons secreting OT, linked to calm and well-being? Gimpl and Fahrenholtz (2001) point to the possibility that a part of the hypothalamic OT reaches the anterior lobe of the pituitary, being able to function as a regulator of stress hormones.

We usually talk about the *amygdala* in the singular, but there is actually a pair of them, each located in one of the brain hemispheres, as already mentioned. The amygdala (or *amygdaloid body*) functions as a kind of sentinel, which sounds a body alert when the organism detects situations linked to important feelings and emotions (Torras et al. 2001). The connections between the BNST and the amygdala are activated in situations of stress; of anxiety linked to conditioned fear (Albuquerque and Silva 2009); of anger, in aggressive encounters or of a sexual nature (Caldwell 2017), as well as in states of long-lasting fear, possibly linked to dysfunctions and psychiatric diseases (Lebow and Chen 2016).

Emotions linked to fear are modulated by the amygdala, including the recognition of facial expressions, and coordinate responses to face threat and danger; they are crucial, therefore, for survival. Finally, they also play a key role in the storage of affective memories (Caldwell 2017).

Here we will open parentheses to clarify a quite common misconception: we often think of stress as a harmful, damaging, and undesirable event, from which we must get rid at any cost. However, the mere existence of the stress response indicates that it has its function, since nature operates economically: no organ or function exists or remains in the organism by chance.

The stress response is an adaptive response of the organism to situations of confrontation, challenge, or threat. In these situations, at a molecular level, specialized cells activate an adaptive pathway called *integrated stress response*, with the aim of restoring cellular homeostasis (Pakos-Zebrucka et al. 2016). When the response is activated in excess, there may be a pathological outcome for health; however, when it is proportional to the stimulus that triggered it, this will be extremely important for overcoming difficulties and for personal development. The adrenergic response to stress will temporarily increase the person's energy and alertness, helping him/her to face the present difficulty. In the behavioral realm, the integrated response to stress mobilizes forces so that the person can face challenges, competition situations, novelties, dangerous situations, and those that represent a life risk.

Finally, the *spinal cord* is a continuation of the *brainstem* and constitutes the lower part of the CNS, where oxytocinergic neurons can also be found (Kandel et al. 2012).

The Oxytocinergic System in the Brain. Signaling Maps. Gene Expression

The oxytocinergic system is constituted by the network formed by OT genes and by AVP and OT receptors (OTR) in the brain and body (Kandel et al. 2012).

The oxytocinergic distribution in the human brain (central) is quite complex and regulates the expression of a wide range of behaviors and physiological functions. Some of its most essential functions are related to social life in general, to maternal care, and to mind-body regulatory effects (Carter 2014). AVP, in turn, besides being an antidiuretic hormone that regulates the body fluids so that the person does not dehydrate, also has an important central role, related to the regulation of sexual behavior, aggression, fear and stress, as well as learning and memory processes (Alescio-Lautier and Soumireu-Mourat 1998).

The formation of oxytocinergic pathways in the brain depends on the presence and location of oxytocin receptor neurons (OTR), toward which OT-producing neurons will project their innervations, connecting to them in a three-dimensional network (Carter 2014). (It is always worth to remember that, if there is no binding to a receptor, an OT neuron cannot express itself).

To uncover the structure of this network, it would initially be necessary to map the OT neurons, their fibers, and OTRs. This map could be the starting point for investigating the forms of OT expression, whether this expression is a behavior, an emotional-affective manifestation, or body, neurochemical, cellular, molecular modifications, etc.

We currently have several methodological possibilities for this type of investigation in the human brain: brain mapping through histological cuts (only possible postmortem), functional imaging (an image is produced while the animal or person is going through a certain type of emotional situation or task, to identify which areas of the brain are being activated). This investigation is more difficult as the target structures nest in deep regions of the brain. Some methodologies are more suitable for animal research, and others, for humans, being comparative studies also important.

The Receptors of OT and AVP

So far, only one type of receptor for OT has been identified, although the hypothesis of having subpopulations of oxytocinergic receptors was not ruled out (Baribeau and Anagnostou 2015).

On the other hand, at least three types of receptors for AVP have been identified: V1a, V1b, and V2, with the peculiarity that *OT binds exclusively to its own receptor*, but *AVP has affinities with the OT receptor*. The OT receptors and the V1a receptors, abundant both in the developing brain and in the adult one, appear to be the main brain receptors in adults (Panksepp 1998).

Recently, Taiwanese neuroscientist Liao and others (2020) investigated the distribution of OT pathways in the CNS of transgenic mice, from stained histological samples of their brains. They noticed a thin web of fibers (axons and dendrites) and cell bodies, with many OT receptors: in the anterior olfactory nucleus, in some cortical regions, in the subcortical regions, mainly in the hypothalamus and brainstem. The cell bodies, in turn, were located exclusively in the hypothalamus and in the bed nucleus of the stria terminalis (BNST), confirming data already obtained. It makes sense that there are so many OTRs in the olfactory region of mice, given the importance that smell has for their survival. In humans, the importance of smell has been relativized with evolution and the expansion of their capabilities, and therefore, its expression, has decreased (Gimelbrant et al. 2004). This is a good example of how evolution can modify gene expression in distinct species.

In another study with human subjects conducted in 2019, the neuroscientist Daniel Quintana, from the University of Oslo, Norway, investigated the distribution of OT genes in the brain from samples of human tissues donated and stored on the Allen Human Brain Atlas platform (http://human.brain-map.org). From these data, they created volumetric expression maps *voxel by voxel*, identifying possible gene interaction pathways, in a comparison of expression patterns of a total of 20,737 genes. They observed that OT mRNA was expressed with great intensity in the *PVN of the hypothalamus*, in the *lateral hypothalamic area*, and in the *supraoptic nucleus*. Another important finding was the evidence of co-expression of OTR mRNA with endogenous opioid, dopaminergic, muscarinic-cholinergic receptors, and genes linked to metabolic regulation, which may be related to social behaviors, psychiatric disorders, and metabolic regulation. Quintana et al.'s study also confirms that the regions with critical OT gene pathways are more abundant in the olfactory and subcortical areas. The authors draw attention to the fact that, of the 20,737 genes studied, the cognitive state maps expression for OTR was among the 0.5% that had stronger relationships with the cognitive state maps of *sexual, motivation, incentive, and anxiety* states, and were statistically significant at ($p < 0.001$). Cognitive state maps for *taste, stress, reward, monetary, fear, and emotional* were among the top 2.5% of all associations with OTR expression, and also significant at ($p < 0.001$) (Quintana et al. 2019, p. 6).

But what does it mean the co-expression of an OTR with another hormone? One of the interactions observed by the authors was that of OT with dopamine (DA), suggesting that they may have a co-expression. Let us hypothesize what would that mean: DA is a hormone that motivates the animal to explore the environment. In this exploration, if it encounters a conspecific (animal of the same species), OT might facilitate the approach and contact without fear between them, and, eventually, a sexual contact. In this case, OT and DA would complement each other's expression, in favor of an eventual generation of life.

As a conclusion of their work, the authors conducted a meta-analysis of studies that used functional magnetic resonance imaging (fMRI) to locate OT genes, concluding that the mapped areas were related to the processing of *anticipatory, appetitive, and aversive cognitive states* (Quintana et al. 2019).

The comparison between the studies of Liao et al. (2020) with Quintana et al. (2019) confirms that, in the brain of mice as well as in humans', OT pathways, in a homologous way, are located with greater density in subcortical regions, where affective-emotional expressions characteristic of these regions are generated (Panksepp and Biven 2012), and in fewer numbers in cortical regions. Could it be that, for the adequate expression of cognitive abilities, the affective-emotional aspects should be less represented?

OT and AVP Are Sexual Hormones

Vasotocin (VAT), the ancestral hormone from which OT and AVP likely derived, was identified in the brains of male frogs and lizards, which are species evolutionarily inferior to mammals. It was found that, in these animals, the secretion of VAT is intimately linked to manifestations of sexual behavior: the infusion of this hormone in their brains immediately provoked courtship sounds and display of sexual behavior. With evolution, VAT dissociated into two hormones in mammals, OT and AVP. Both are synthesized in the brains of males and females, although in different proportions: OT is produced on a larger scale by the female brain, and AVP, by the male brain (Panksepp 1998).

This difference points to a specific hormonal specialization for each sex, contributing to the development of different expressions according to gender and to the complementary relationship between them, as already mentioned (Carter 1998). Therefore, at least regarding these hormones and their expression, the brains of male and female mammals are different, or dimorphic (MacDonald and Feifel 2013).

And how would these sexual hormonal differences phenotypically express themselves? OT is a hormone that quiets and calms the female, making her more receptive to the male's sexual approach. AVP, on the other hand, stimulates the male's sexual insistence in approaching the female, expressing behaviors of harassment, courtship, and consummation of sexual act. In addition, it can express itself through aggressive behaviors, to protect its offspring and the female. This applies to rodents and primates. In humans, these issues are more complex, because, in addition to instincts, humans are guided by several layers of cultural influences: from the region of birth, from the group of friends, from the family group. Still, there are a series of studies that emphasize the influence of OT in the relationship with a different sex in humans (Ditzen et al. 2009).

In terms of expression of sexuality, the release of OT is indispensable for penile erection and male and female orgasm, in addition to creating an atmosphere of approach without fear and positive affection (Porges 2011).

OT Is an Evolutionary Hormone

There is plenty of evidence that OT is an evolutionary hormone, and that the evolution of the brain apparently is linked to this hormone.

The American neuroscientist C. Sue Carter, a pioneering scholar of central OT, argues in a 2014 study that high levels of sociability, complex social interactions, and sophisticated social constructions could not have been developed without the support of OT articulated with AVP, acting on sensory, visceral, and metabolic regulation. On the other hand, still according to her, social behavior would also have been essential for evolution, via multiple genetic and physiological substrates (Carter 2014). In a 2009 study, American neuroscientist Ralph Adolphs also emphasized that the human nervous system is a product of adaptation for social life (Adolphs 2010). This hypothesis is interesting because it suggests a circular mechanism of influences between social and evolutionary mechanisms. Ebstein et al. (2012) add that the signaling pathways of OT suffer the influence of genetic variation, being epigenetically refined by social experiences and exposure to hormones.

C. Sue Carter adds that, especially concerning specializations in socialization and consequent structural modifications in the genome, the oxytocinergic pathways, formed by OT, AVP, and their receptors, could be at the center of the physiological and genetic systems that allowed the evolution of the human NS, until reaching the expression of human sociability as we know it today (Carter 2014). A genetic regulation allows for the growth of the neocortex in the human brain and maintains the blood supply to the cortex (Carter 2014).

Interestingly, unlike most hormones, the regulation of OT occurs by positive feedback. This means that the production of OT leads to effects that stimulate even more the release of this hormone. On the other hand, AVP is regulated in the opposite way, with negative feedback, that is, after the production of a certain amount of hormone, the process is interrupted. Therefore, in an evolutionary scenario, OT would have the potential to increasingly irrigate social processes (Uvnäs-Moberg 1998a, b).

By supporting maternal behaviors in females, OT indirectly also aids evolution, as this behavior, when occurring within normal parameters, directly influences the development of the baby's nervous system.

Pedersen et al., in a 2014 study, proposed that the effects of OT in the CNS were responsible for crucial achievements related to social behavior during the evolution of placental mammals (placenta-bearing), which constitute most living mammal groups. The placenta itself would be an evolutionary improvement, as it allows for the exchange of nutrients and excretions through maternal blood, as well as providing a longer period of fetal maturation inside the mother's womb, greatly increasing the chances of offspring survival. In addition, OT also activates maternal behavior of taking care of the offspring, which, in addition to breastfeeding, includes the protection of offspring and adds affective aspects, thus reducing fear and anxiety in the offspring, facilitating the receipt of care and breastfeeding.

The development of sociability, therefore, is linked to the activity of OT, and is an important resource to improve the chances of human survival and its quality of life, essential for the development and regulation of the organism.

OT Is a Social, Anti-Panic, and Immunological Hormone

OT and AVP organize and modulate not only sexual functioning but also social behavior in mammals.

The release of OT, characteristic of affective social contacts and of support to others, is strongly associated with anti-stress effects, since social contact is a powerful attenuator and regulator of stimuli causing suffering, unpleasant and nociceptive feelings (Uvnäs-Moberg 1998b; Schore and Schore 2008). In a 2009 study, the German neuroscientist Inga Neumann pointed out that social contact can have an immunological function because of the stress attenuation produced by the release of OT (Neumann 2009). This anti-stress social support effect was observed in small babies, and in social contact with sick people, which is why ill people benefit emotionally and recover faster when they receive social support delivered by friends, relatives, and hospital staff (affective regulation) (Hendryx et al. 2009).

In the event of an early separation between the components of the mother-offspring dyad, panic behaviors were observed in the offspring, with great motor agitation and vocalizations characteristic of suffering, mobilized by the activation of other specific neural circuits (Panksepp 1998; Grippo et al. 2007). In humans, early separation between the baby and the mother causes serious psychoneurobiological problems in the child, increasing morbidity and mortality rates. This situation can be reversed if the mother and the child are reunited within a certain critical period of time. More recently, there have been disagreements about these findings (Spitz 1945; Myers 1984; Howard et al. 2011).

In 1998, Nelson and Panksepp observed that, in rats, there seemed to be an attenuation of separation stress when *OT, endogenous opioids (OP), estrogen, and epinephrine* were released by social stimuli, modulating affiliative behaviors, in a way not yet fully clarified by science, as confirmed by Jaak Panksepp in a written communication by e-mail, on August 25, 2015.

Evolutionary social repertoires bring important benefits to mammals, both individually and as members of groups, as, for example, in the development of intragroup loyalty versus intergroup aggression (De Dreu and Gross 2018); conspecific cooperation in family baby care (König 1997) and formation of affiliative and community groups (Feldman 2012).

OT Is a Regenerative and Analgesic Hormone

In addition to its social-immunological interface, the release of OT is also associated with physiological effects of shortening the period of healing of cuts and wounds (Uvnäs-Moberg et al. 2005; Poutahidis et al. 2013).

OT also plays a role in the modulation of painful phenomena, reducing the intensity of pain. This is due, on the one hand, to the attenuation of stress when the person receives social and emotional support (Wang et al. 2013), and, on the other hand, to the association of the oxytocinergic system with endogenous opioids (Yang et al. 2011).

There is evidence of the participation of OT in both physical pain and the emotional pain of separation, if one can, in fact, speak of a duality between emotional and nociceptive pain, since, in neuroanatomical terms, the areas activated by these two types of pain are partially overlapped, and, apparently, the brain area for emotional pain is an evolutionary specialization of the nociceptive area (Eisenberger 2012).

OT Is a Regulatory Hormone

OT seems to have a relevant role in digestion, sleep, and reproduction functions (Porges 1998); in the regulation of the HPA axis (Carter 1998) and in behavioral, neuro-endocrine-immunological, and autonomous aspects (Schore 2001), especially in its connections with the inhibitory vagal system, which slows down the heart rate (vagal brake), contributing to the homeostasis of the being (Porges 1998).

Regarding the anxiolytic aspects of OT, the central or subcutaneous infusion of OT in rats seemed to provoke anxiolytic effects at low doses and sedative effects at higher doses, indicating that OT may have *dose-dependent* effects (Uvnäs-Moberg 1998a). In a 1998 study, Carter suggested that the positive effects of sedation, calm, and anti-stress would be caused by the activation of secondary mechanisms, such as the increase of endogenous opioid activity. MacDonald (2013) observed that, in situations of aversive attachment, there may be a negative impact in the stress system.

An interesting aspect of OT, widely explored by Uvnäs-Moberg in 1994, and later, in association with her collaborators Linda Handlin and Maria Petersson (Uvnäs-Moberg et al. 2015), is the fact that its release is easily activated by *pleasant natural sensory stimuli*, such as massage; certain aromas; certain music; rocking movement; pressure on the body; warm temperature; and skin touch, that is, stimuli that characteristically occur in mothering. According to Uvnäs-Moberg, OT remains in the body for many hours after its release/administration, and its release, even in very small quantities, can produce important and prolonged physiological and behavioral effects (Uvnäs-Moberg 2003).

OT and Libido

We will close this chapter with a neuropsychoanalytic observation: as neuroscience identifies that *OT—a mammalian hormone of sexual phylogeny*—plays a fundamental role in sexual, parental, and social relations, and that, in humans, it has *species-specific aspects,* we observe a similarity between the role of this hormone and the conception of sexuality in Freud's psychoanalysis, which, according to him, would be present in object relations in the form of a vital energy, which he called *libido* (Freud 1915, 1922).

To create his conception of libido, Freud used the same term from erotic literature which, in Latin, means *will, desire*. The Freudian libido would be "the substrate for the transformation of sexual impulses" (Laplanche and Pontalis 1973).

According to Freud:

Libido is an expression taken from the theory of affectivity. This is what we call this energy, considered quantitative, although it is not actually measurable,—of the drives that refer to everything that we can understand under the name of love. (p. 344)

The French psychoanalysts Jean Laplanche and Jean-Bertrand Pontalis Lefebvre, authors of the classic *Vocabulary of Psychoanalysis* (1973), a renowned reference work in psychoanalysis, observe that the concept of libido is "clearly distinct from the concept of somatic excitation [...] The sexual drive is located on the psychosomatic limit [...] and is clearly distinct from the concept of somatic excitation [...] and the Freudian libido would be the energy of this drive" (Laplanche and Pontalis 1973, p. 344). It is possible that the behavioral-scientific knowledge we have today about the social/sexual roles of OT may stimulate and even contribute to the adventure of a rereading of the Freudian libidinal theory, offering a physiological support to the notion of sexual drive as an internal pressure. Still, according to Laplanche and Pontalis, "it operates in a much broader field than that of sexual activities in the current sense of the term [...]. Their satisfaction modalities [would be] likely to accompany the most diverse activities on which they rely [...]. Freud postulates the existence of a unique energy in the vicissitudes of the sexual drive: the libido" (Laplanche and Pontalis 1973, p. 518).

References

Adolphs R (2010) Conceptual challenges and directions for social neuroscience. Neuron 65(6):752–767

Albuquerque FDS, Silva RH (2009) A amígdala e a tênue fronteira entre memória e emoção. Rev Psiquiatr Rio Gd Sul 31(Supl. 3):1–18

Alescio-Lautier B, Soumireu-Mourat B (1998) Role of vasopressin in learning and memory in the hippocampus. Prog Brain Res 119:501–521

Avery SN, Clauss JA, Blackford JU (2016) The human BNST: functional role in anxiety and addiction. Neuropsychopharmacology 41(1):126–141

Baribeau DA, Anagnostou E (2015) Oxytocin and vasopressin: linking pituitary neuropeptides and their receptors to social neurocircuits. Front Neurosci 24(9):335

Caldwell HK (2017) Oxytocin and vasopressin: powerful regulators of social behavior. Neuroscientist 23(5):517–528

Carter CS (1998) Neuroendocrine perspectives on social attachment and love. Psychoneuroendocrinology 23(8):779–818

Carter CS (2014) Oxytocin pathways and the evolution of human behavior. Annu Rev Psychol 65:17–39

De Dreu CKW, Gross J (2018) Revisiting the form and function of conflict: neurobiological, psychological, and cultural mechanisms for attack and defense within and between groups. Behav Brain Sci 42:e116

Ditzen B, Schaer M, Gabriel B, Bodenmann G, Ehlert U, Heinrichs M (2009) Intranasal oxytocin increases positive communication and reduces cortisol levels during couple conflict. Biol Psychiatry 65(9):728–731

Ebstein RP, Knafo A, Mankuta D, Chew SH, Lai PS (2012) The contributions of oxytocin and vasopressin pathway genes to human behavior. Horm Behav 61(3):359–379

Eisenberger NI (2012) The neural bases of social pain: evidence for shared representations with physical pain. Psychosom Med 74(2):126–135

Esperidião-Antonio V, Majeski-Colombo M, Toledo-Monteverde D, Moraes-Martins G, Fernandes JJ, Bauchiglioni de Assis M, Montenegro S, Siqueira-Batista R (2017) Neurobiology of emotions: an update. Int Rev Psychiatry 29(3):293–307

Feldman R (2012) Oxytocin and social affiliation in humans. Horm Behav 61(3):380–391

Freud S (1915) A metamorfose da puberdade. In: Obras Completas, vol 2, 3a edn. Ballesteros, Madrid

Freud S (1922) Psicanálise e teoria da libido. In: Obras Completas, vol 3, 3a edn. Ballesteros, Madrid

Fuxe K, Borroto-Escuela DO, Romero-Fernandez W, Ciruela F, Manger P, Leo G, Díaz-Cabiale Z, Agnati LF (2012) On the role of volume transmission and receptor-receptor interactions in social behaviour: focus on central catecholamine and oxytocin neurons. Brain Res 1476:119–131

Gimelbrant AA, Skaletsky H, Chess A (2004) Selective pressures on the olfactory receptor repertoire since the human-chimpanzee divergence. Proc Natl Acad Sci USA 101(24):9019–9022

Gimpl G, Fahrenholz F (2001) The oxytocin receptor system: structure, function, and regulation. Physiol Rev 81(2):629–683

Grippo AJ, Gerena D, Huang J, Kumar N, Shah M, Ughreja R, Carter CS (2007) Social isolation induces behavioral and neuroendocrine disturbances relevant to depression in female and male prairie voles. Psychoneuroendocrinology 32(8–10):966–980

Hendryx M, Green CA, Perrin NA (2009) Social support, activities, and recovery from serious mental illness: STARS study findings. J Behav Health Serv Res 36(3):320–329

Hofman MA, Swaab DF (1992) The human hypothalamus: comparative morphometry and photoperiodic influences. Prog Brain Res 93:133–147

Howard K, Martin A, Berlin LJ, Brooks-Gunn J (2011) Early mother-child separation, parenting, and child well-being in early head start families. Attach Hum Dev 13(1):5–26

Kandel ER, Schwartz JH, Jessell TM, Siegelbaum SA, Hudspeth AJ (2012) Principles of neural science, 5th edn. McGraw Hill

König B (1997) Cooperative care of young in mammals. Naturwissenschaften 84(3):95–104

Laplanche J, Pontalis JB (1973) The language of psycho-analysis (Trans. Donald Nicholson-Smith). W. W. Norton

Lebow MA, Chen A (2016) Overshadowed by the amygdala: the bed nucleus of the stria terminalis emerges as key to psychiatric disorders. Mol Psychiatry 21(4):450–463

Liao PY, Chiu YM, Yu JH, Chen SK (2020) Mapping central projection of oxytocin neurons in unmated mice using Cre and alkaline phosphatase reporter. Front Neuroanat 14:559402

Macdonald KS (2013) Sex, receptors, and attachment: a review of individual factors influencing response to oxytocin. Front Neurosci 6:194

Macdonald K, Feifel D (2013) Helping oxytocin deliver: considerations in the development of oxytocin-based therapeutics for brain disorders. Front Neurosci 7:35

Magon N, Kalra S (2011) The orgasmic history of oxytocin: love, lust, and labor. Indian. J Endocrinol Metab 15(Suppl 3):S156–S161

References

Mahlmann S, Meyerhof W, Hausmann H, Heierhorst J, Schönrock C, Zwiers H, Lederis K, Richter D (1994) Structure, function, and phylogeny of [Arg8]vasotocin receptors from teleost fish and toad. Proc Natl Acad Sci USA 91(4):1342–1345

Moore KL, Persaud TVN, Torchia MG (eds) (2020) The developing human: clinically oriented embryology, 11th edn. Elsevier, Berkeley

Myers BJ (1984) Mother-infant bonding: the status of this critical-period hypothesis. Dev Rev 4(3):240–274

Nelson EE, Panksepp J (1998) Brain substrates of infant-mother attachment: contributions of opioids, oxytocin, and norepinephrine. Neurosci Biobehav Rev 22(3):437–452

Neumann ID (2009) The advantage of social living: brain neuropeptides mediate the beneficial consequences of sex and motherhood. Front Neuroendocrinol 30(4):483–496

Pakos-Zebrucka K, Koryga I, Mnich K, Ljujic M, Samali A, Gorman AM (2016) The integrated stress response. EMBO Rep 17(10):1374–1395

Panksepp J (1998) Affective neuroscience: the foundations of human and animal emotions. Oxford University Press, Oxford

Panksepp J, Biven L (2012) The archaeology of mind: neuroevolutionary origins of human emotion. WW Norton & Company, New York

Pedersen CA (2014) Schizophrenia and alcohol dependence: diverse clinical effects of oxytocin and their evolutionary origins. Brain Res 1580:102–123

Porges SW (1998) Love: an emergent property of the mammalian autonomic nervous system. Psychoneuroendocrinology 23(8):837–861

Porges SW (2011) The polyvagal theory: neurophysiological foundations of emotions, attachment, communication, and self-regulation (Norton series on interpersonal neurobiology). WW Norton & Company, New York

Poutahidis T, Kearney SM, Levkovich T, Qi P, Varian BJ, Lakritz JR, Ibrahim YM, Chatzigiagkos A, Alm EJ, Erdman SE (2013) Microbial symbionts accelerate wound healing via the neuropeptide hormone oxytocin. PLoS One 8(10):e78898

Quintana DS, Rokicki J, van der Meer D, Alnæs D, Kaufmann T, Córdova-Palomera A, Dieset I, Andreassen OA, Westlye LT (2019) Oxytocin pathway gene networks in the human brain. Nat Commun 10(1):668

Schore AN (2001) The effects of a secure attachment relationship on right brain development, affect regulation, and infant mental health. Infant Ment Health J 22(1–2):7–66

Schore JR, Schore AN (2008) Modern attachment theory: the central role of affect regulation in development and treatment. Clin Soc Work J 36(1):9–20

Spitz RA (1945) Hospitalism; an inquiry into the genesis of psychiatric conditions in early childhood. Psychoanal Study Child 1:53–74

Torras M, Portell I, Morgado I (2001) La amígdala: implicaciones funcionales. Rev Neurol 33(5):471–476

Uvnäs-Moberg K (1996) Neuroendocrinology of the mother-child interaction. Trends Endocrinol Metab 7(4):126–131

Uvnäs-Moberg K (1998a) Antistress pattern induced by oxytocin. News Physiol Sci 13:22–25

Uvnäs-Moberg K (1998b) Oxytocin may mediate the benefits of positive social interaction and emotions. Psychoneuroendocrinology 23(8):819–835

Uvnäs-Moberg K (2003) The oxytocin factor: tapping the hormone of calm, love and healing. Boston, Massachussets. Da Capo Press

Uvnäs-Moberg K, Arn I, Magnusson D (2005) The psychobiology of emotion: the role of the oxytocinergic system. Int J Behav Med 12(2):59–65

Uvnäs-Moberg K, Handlin L, Petersson M (2015) Self-soothing behaviors with particular reference to oxytocin release induced by non-noxious sensory stimulation. Front Psychol 5:1529

Wang YL, Yuan Y, Yang J, Wang CH, Pan YJ, Lu L, Wu YQ, Wang DX, Lv LX, Li RR, Xue L, Wang XH, Bi JW, Liu XF, Qian YN, Deng ZK, Zhang ZJ, Zhai XH, Zhou XJ, Wang GL, Zhai JX, Liu WY (2013) The interaction between the oxytocin and pain modulation in headache patients. Neuropeptides 47(2):93–97

Yang J, Liang JY, Li P, Pan YJ, Qiu PY, Zhang J, Hao F, Wang DX (2011) Oxytocin in the periaqueductal gray participates in pain modulation in the rat by influencing endogenous opiate peptides. Peptides 32(6):1255–1261

Chapter 5
Attachment

Development of the Concept of AT: Bowlby, Spitz, Winnicott

The term *attachment* was coined by the English psychologist, psychiatrist, and psychoanalyst John Bowlby (1907–1990), to designate the loving bond developed between mother and baby, within the context of his attachment theory, developed from 1958 (Bowlby 1969).

Two other contemporary professionals of Bowlby, René Spitz (1887–1974), and Donald Winnicott (1896–1971), also contributed with their own research, to a broader understanding of the mother-baby relationship, as well as the consequences of a separation or definitive rupture of this dyad (Spitz 1945; Winnicott 1948).

Bowlby, in the same way as the English psychoanalyst Winnicott, concluded in his clinical work, the connection between juvenile delinquency and an inadequate and/or insufficient contact between mother and child (Spitz 1945; Winnicott 1948).

The first studies in human AT were mainly related to its absence; to situations where its development was not possible; or where it did not occur sufficiently and adequately, due to early separations, loss of parents, or the relationship with an absent, depressed, helpless, rejecting mother, or that was unable to decode the needs and rhythm of her baby (Spitz 1945; Winnicott 1948; Bowlby 1958); or, still, due to unfavorable epigenetic situations (Champagne and Curley 2009).

Handling, Holding, and the Good Enough Mother

At birth, the human baby, due to its extreme physiological and psychological immaturity, cannot survive and develop competent social characteristics, unless cared for by another more developed being of its species (Hong and Park 2012). This period of dependence is much longer in humans than in most mammals. The prolonged and tiring work of providing physical, psychic, and emotional support to a baby, which Winnicott called *holding* (Winnicott 1962), as well as its adequate *care*, named by him *handling* (Winnicott 1970), will require a lot of time, dedication, sensitivity, and patience from the caregiver, who, generally, is the mother.

In favorable contexts, evolution seems to have selected strategies to favor the emergence of a special positive bond between mother and baby, ensuring that this AT bond and childcare are maintained for a long time. Generally, the mother demonstrates an aesthetic and narcissistic appreciation for her baby, finding him beautiful, with a pleasant smell, showing herself enchanted with his performances and babbling, and by what she understands as his progress, beauty, and intelligence.

Although the mother often feels exhausted and irritated with how much work the baby takes, she generally manages to contain herself and be adequate, or "good enough," in relation to her child, as proposed by Winnicott (1953). On his side, the baby is also enchanted and attached to his mother; tries to get her attention, tries to cling to her, cries in her absence or departure, needs her to understand his needs, and rejoices with her return, if the separation is not excessively prolonged (Spitz 1945; Bowlby 1969; Winnicott 2005).

A few years before Winnicott developed his theory, the Viennese psychoanalyst based in the United States, René Spitz, had already realized the dire consequences of early mother-baby separation, while working with children who had been taken away from their parents, and institutionalized in a place near New York, during the Second World War. Spitz (1945) called the resulting psychological situation, of *hospitalism syndrome*. In this case, the babies, although cared for and fed regularly, but without affection and without being held, showed impaired physical development: they had no appetite, did not gain weight, and then lost interest in relationships. If the reunion with the parents did not occur within a critical period, which he estimated between 3 and 5 months, this situation would become irreversible and linked to a high rate of morbidity and mortality (Spitz 1945).

Bowlby, who, in addition to being a psychologist, psychiatrist, and psychoanalyst, was also a neuroscientist *avant la lettre,* achieved an AT conception by integrating psychological and psychoanalytic, biological, and physiological components, detailed in the trilogy *Attachment and Loss* (1969). He began to elaborate his Theory of Attachment while still young, from an experience conducted with institutionalized adolescents from dysfunctional families, when he noticed a prevalence of the development of *juvenile delinquency* among the youth who did not have satisfying relationships with their mothers (Bowlby 1984).

In 1937, Bowlby joined the British Psychoanalytical Association, and tried to develop his theory within this context. However, some colleagues interpreted that his proposals violated certain basic principles of psychoanalysis. This created a great controversy, which culminated in Bowlby's expulsion from the Association (Fonagy et al. 2018). Resentments aside, this rupture left Bowlby free to drink from other sources of knowledge, with ample freedom to investigate them, which resulted in original and essential contributions to the development of his Theory of Attachment.

Relevant Influences in the Development of Bowlby's Theory of Attachment

In the elaboration of his theory, Bowlby received numerous influences from other scientists. A first and fundamental influence was the *Darwinian evolutionary epistemology*, as Bowlby himself verbalized it (Bowlby 1982). According to evolutionary principles, those strategies and behaviors that best ensure the survival of the species are preserved and perfected by evolution, favoring the fittest beings, which have better chances of survival. This process takes place within the evolutionary constraints of each species.

The *scientific ethology* was also an important influence that occurred thanks to the cooperation between Bowlby and the ethologists Konrad Lorenz, Nikolaas Tinbergen, and Robert Hinde. The concepts of the ethologists, especially that of *imprinting* (Hess 1959), helped him to create a *biological interface* of the Theory of Attachment with ethology. This contact also led Bowlby to adopt a *scientific referral* in his works.

Another study that reinforced Bowlby to scientifically ground AT was the one that investigated the nature of love in baby rhesus monkeys, conducted by psychologist Harry Harlow, a study situated in the interfaces of ethology, biology, and animal psychology.

Other relevant influences for Bowlby's work were *cybernetics and information processing* (Bretherton 1992), and Spitz's data over the *prolonged mother-baby separation* (Spitz 1945).

Finally, it is necessary to mention the influence of French epistemologist Jean Piaget, with whose theory, *genetic epistemology*, Bowlby professed several affinities (Pallini and Barcaccia 2014).

Interestingly, Bowlby ended up working in a context of great complexity, using elements of comparative psychology, anticipating the epistemological practices of contemporary neuroscience.

Over time, Bowlby moved away from psychoanalytic epistemology, increasingly adopting scientific references, and developing interfaces with other areas of knowledge. In his words:

> I have been developing a paradigm that, whilst incorporating much psychoanalytic thinking, adopting several principles that derive from [...] ethology and control theory. [...] Whilst its concepts are psychological [...] they are compatible with those of neurophysiology and developmental psychology, and they can meet the ordinary requirements of a scientific discipline. (Bowlby 1969 p. 38)

While describing AT, Bowlby says that it is a basic instinctive biological mechanism, a "lasting bond between human beings" (Bowlby 1969), and associates it with *love*, highlighting the intense emotions linked to the formation, maintenance, rupture, and rapprochement in mother-baby affective relationships (Bowlby 1980).

Bowlby will differentiate AT from the instincts of feeding and sexual drive, emphasizing that "food and feeding are held to play no more than a minor part in (the baby) development" (1969 p. 180). According to him, AT performs a *homeostatic function* in the mother-baby dyad, along with other behavioral systems: "the child's tie to his mother is a product of the activity of several behavioral systems that have *proximity to mother* as a predictable outcome" (1958 p, 179). These systems would arise from the interaction with the mother, according to the genetic programming characteristic of the human species.

Bowlby, Ainsworth, and the Internal Working Models

Bowlby and his collaborator, Mary Ainsworth, expanded the Theory of Attachment by creating a psychic construct they called *internal working model* (IWM). These models would be formed in the child's mind, under the form of mental/emotional representations of self and others, according to the lived experiences (Steele 2003). Some followers of Bowlby proposed that the development of IWMs in the baby would give rise, in him, to a *theory of mind*, capable of *predicting mental and behavioral functioning*, in oneself and others, thus transforming the human Theory of Attachment in a higher order representation (Fonagy and Target 1997). Therefore, although the AT presents similar mechanisms in different people, it will bring an individualized outcome to each of them, depending on their experiences and personal genetics.

Just as he was influenced by various areas of knowledge, Bowlby also influenced many other areas, such as animal ethology (Hess and Petrovich 2000); affective neuroscience (Nelson and Panksepp 1998); educational psychology (Kennedy and Kennedy 2004); and, more recently, evolutionary, developmental, and regulatory neuroscience (Schore 1994; Carter 1998; Porges 2011). Years later, the British Psychoanalytic Association recognized the importance of Bowlby and his theory, reviewed its position, and incorporated it into psychoanalytic theory (Sroufe 1983; Diamond 1999; Fonagy et al. 2002), emphasizing the polarity between AT and separation (Blatt and Levy 2003) and studies with the *Strange Situation Test*, developed by Ainsworth et al. (1978). According to this test, the possible responses of a child, when left alone with a stranger after the mother leaves the room, can be: (a) those typical of *secure attachment*; (b) those of *anxious-resistant insecure attachment*; (c) those of *anxious-avoidant insecure attachments*; and (d) those of *disorganized-disoriented attachment*, this last one later identified by Main and Solomon (1986). The test would allow the identification of the child's mental construction of the attachment dynamics.

We mention here some important characteristics of AT:

- "Intense and long-lasting emotional link […] includes the responses of rooting, sucking, and holding […]. Intense motivational and emotional load […], tenacity" […] (Mason and Mendoza 1998).
- "Attempt of an individual (generally the son) in keeping close of a specific other (generally the mother), exhibiting suffering when the separation or loss of the other occurs, and trying to re-establish proximity after the separation" (Mendoza and Mason 1997).

- "Has the function of facilitating reproduction, provide feelings of safety, and reduce feelings of stress or anxiety" (Carter 1998).
- It would be a selective social or emotional bond (Ainsworth et al. 1978; Bowlby 1969, 2002; Winnicott 1990).

The historical process of the development of Bowlby's Theory of Attachment resulted in the accumulation of a large multidisciplinary and in evolution database, not necessarily systemized into a cohesive whole. This multidisciplinary approach also gave rise to controversies, caused mainly by methodological and epistemological issues, among which the very concept of AT: What would it be? What are its functions? What is its evolutionary role? Some scientists even questioned whether the very concept of AT would not be an attempt to reify something that could simply be defined in behavioral terms (Gubernick 1981). Currently, neuroscientific discoveries seem to leave no doubt about the *multifaceted and transdisciplinary aspects* of AT.

As AT can manifest itself in an unidirectional or reciprocal way in animal species, especially in mammals, ethological studies investigated its various vector possibilities: preference of the offspring for the mother over another caregiver, and preference of the mother for caring for her offspring over offspring from other litters; the search for proximity on the part of the offspring; responses to separation for short and long periods in the female and the offspring; and responses to reunion between them. From these criteria, it was observed, for example, that many rodents do not exhibit AT behavior, but monkeys do, with large variations between species, and ungulates (animals that have hooves) also do, with variation in sensory modalities (vision, smell, hearing) that mediate this phenomenon (Gubernick 2013).

The subsequent development of neuroscience allowed us to understand that the AT does not occur as an independent behavioral pattern, but as *an amalgamated and interdependent part of psychological and neurobiological systems*, and from influences from the external environment; in humans, it largely results from the child's maturation processes, from evolutionary, cultural, and influential social groups pressures, with the current emphasis in virtual modalities. The manifested AT will result from phylogenetic,[1] genetic,[2] ontogenetic,[3] and epigenetic[4] components (Beltrame 2011).

Panksepp (1998) emphasizes that the critical information about AT in human development came from animal experiments, and that the brain processes underlying AT, gathered in investigations with animal models, formed the basis for the initial understanding of these processes in humans. This possibility derives from the neuroanatomic homology of the limbic brain areas in lower mammals and humans, notably well preserved in the evolution of the mammalian brain, responsible for the emergence of analogous affective-emotional manifestations.

---

[1] Relative to the species.
[2] Relative to the genome.
[3] Relative to the development of the being after birth.
[4] Relative to the modifications of the genome by the environment, without alteration of the DNA.

Discoveries about the human nature from comparative neuroscience have initially caused a reasonable social commotion, as it relativizes the supposed superiority of human emotional manifestations in relation to "inferior" species. They seem to have subsequently influenced the search for more ethical ways of conducting experiments with animals, and even human-animal relations.

Attachment in Lower Mammals and in Humans

In mammals in general, the AT process in childhood is closely related to survival and good development of offspring. Under normal conditions, the mother's AT is mainly manifested through the release of OT (Carter 1998), contributing to the mothering behaviors: *breastfeeding, cleaning and comfort, removal of painful stimuli, touch (licking, grooming), and pressure (lying on the offspring),* with specific characteristics for each species (Mendoza and Mason 1997).

These findings corroborate Harlow's (1959) earlier findings about the preference of monkey babies for a soft "substitute mother" more pleasant to the touch (made of plush), in comparison with a "substitute mother" made of wire. Mammals, including humans, feel comfortable, calm down, and may sleep when placed in a sensorially comfortable, warm, and sheltered environment. In some phylogenetically inferior species (birds, for example) this also occurs (Panksepp 1998).

Observations of human babies indicate that, under normal conditions, AT develops bidirectionally between mother and baby (Panksepp et al. 1978), having high survival value, given the prematurity of the baby at birth and the long time needed for care until its functional emancipation. Besides maternal behavior already mentioned, in the human case, social and loving aspects typical from the species are added, that is, positive affective stimulation between mother and baby through gaze, smell, smiling, prosodic speech, singing, and playing (Trevharten 2003). During the development of AT, an adequate mother would have the primary function of modulating her baby's affects, whose still poorly myelinated nervous system has a deficient functioning, interpreting, attributing meaning and naming his emotions, helping him to face difficult and stressful situations, and calming him down when necessary, as, for a baby, a world full of novelties and unknown situations is very stressful (Winnicott 1984). This process constitutes the *modulation of affect* (Schore 1994; Fonagy et al. 2002) and is only possible when an experienced adult can assist in the development of a baby still in the process of *becoming*, while not yet fully ready for life in the outside.

As an evolutionary contribution, human facial features have changed over millennia, gaining greater expressiveness, with the displacement of the eyes to the front of the face, and the connection of facial muscles with the viscera, via the vagus nerve. The mutuality of smiling and looking seems to be able to convey primary emotional states with great accuracy and intensity, increasing the possibilities of communication and loving bonding. With the evolution of the human ear (musculature, bone structure, and lateral localization of the ear) and the vocal apparatus, vocalizations and prosody became more sophisticated, playing a key role in primordial social and loving exchanges (Porges 2011).

A Baby's Reaction to Its Mother's Care and the Formation of the AT

Besides the care of the mother, AT development also seems to depend on *genetic aspects of the baby's temperament*, as well as those of the mother, which will influence its perception/interpretation of the care received (Steele 2003). Those who choose the clinic as a professional choice know the clients can construct idealized images of their parents in which they exaggerate either their positive qualities or the negative ones. There is then an interpretive bias in the mother's conception in relation to the child, and vice versa, which can be more, or less, close to their real qualities.

When there is an adequate style of mothering, homeostatic mechanisms will develop as sensations of calm, well-being, and security in the mother-baby dyad. In addition, the *adequate mother* will function as a buffer of the aggressive impact of negative stimuli, mediating the baby's relationship with the world, and balancing its physiological and affective-emotional functioning. Later, she should also act in the development of its social behavior, and the more cognitive aspects regarding the environment. Finally, she should be able to help the child to emancipate from her, the mother, and gradually gain autonomy (Winnicott 1984; Schore and Schore 2008; Porges 2011).

In other words, if the regulation of the baby by the mother succeeds through a type of learning that is affective-emotional as well as cognitive, this will lead to mechanisms of self-regulation. That would be, in psychoanalytic terms, the equivalent to the introjection of the good mother in the baby (Klein 2013).

The gradual emancipation between the members of the dyad throughout the AT process seems to lead to the emergence of relatively stable forms of functioning in the baby, which can last a lifetime, serving as a model for future relationships (Fonagy et al. 2002). Indicators of a well-developed AT would be (a) the capacity of identifying emotional availability in the other, (b) the probability of receiving emotional support in stress situations, and (c) the appropriate way of interacting with the others (Bowlby 1969).

Waters et al. (2000) suggested that, throughout cognitive development, the sensory-motor representations of experiences lived on a secure base could give rise to increasingly complex mental representations. Currently, the neuroscientific construct of AT integrates evolutionary, developmental, ethological, psycho-affective, instinctive, neuro-immuno-endocrinological, genetic, and molecular aspects, establishing bridges between the psychic and the physical aspects in the apprehension of oneself and the others.

In the clinic of adults, the consequences of the mother-baby relationship may appear in a complex and paradoxical way—sometimes, as a result of a dysfunctional family dynamics, the person tends to seek relationships that are doomed to failure, that will bring pain and suffering, but that, nevertheless, reproduce their original relationships. What would be the meaning of this? Although they are deviant behaviors if compared to the healthy norm, they make sense to the person, because, in terms of psychic economy, they are familiar to him, and, at times, the only ones he knows; they emerge from primitive neural and endocrinological circuits, configuring a "neural signature" unique to him (Saitovitch et al. 2019).

Attachment and Love

In 1959, the psychologist Harry Harlow used the term "love," when studying the relationship between rhesus monkey infants and "surrogate mothers," constructed with wire, in a conical shape and rudimentary face, or covered with plush, with a face more like that of a monkey. In the outcome, the infants clung most of the time to the plush mother, disregarding the wire mother even when a bottle was attached to it, suggesting that the search for certain positive sensory sensations could take precedence over hunger. The author also emphasizes the face similar to the species as a factor for the preference (Harlow 1959).

Bowlby will also use the term *love* when referring to children with destabilizing family experiences, in the second edition of his report on the subject, prepared for the World Health Organization (WHO), and entitled as *Childcare and the growth of love* (Bowlby et al. 1965).

The use of the word "love" may sound familiar to psychology and psychoanalysis, but it may be less usual in the biological sciences. Even so, in 1998, amid the Decade of the Brain, the journal *Psychoneuroendocrinology,* from Elsevier, published a special edition organized by Kerstin Uvnäs-Moberg and C. Sue Carter, whose title was: *Is there a neurobiology of love?* This question had already arisen in 1996, in a meeting of scientists in Stockholm, which aimed at examining the concept of love from various points of view: biological, evolutionary, behavioral, and neuro-endocrinological (Uvnäs-Moberg and Carter 1998). The scope of this study was thus expanded beyond the usual observable behaviors, delving into the neuroanatomical and hormonal substrates corresponding to what the authors conceptualized as love, whose biological roots would be fundamentally linked to reproductive, maternal, and sexual behaviors; and, secondarily, to positive social behavior. At the endocrinological level, OT appeared as a fundamental component (Uvnäs-Moberg and Carter 1998).

What would be the relationship proposed by neuroscientists between AT and love? Crews (1998) suggested that AT, or affiliation, would be an essential component of love, since, evolutionarily, it would be indispensable to social organization.

Porges (1998) classified the results of love into categories of reproduction, cooperation for survival, transmission of culture, and pleasure and ecstasy, also noting that, although love is considered a human emotion, several neurobiological processes linked to the experience and expression of love in humans *are shared with other mammals*. In an enlightening way, the author titled his article as *Love: an emerging property of the mammalian autonomic nervous system*. In the literature, the terms *AT, bonds, love,* and *positive social feelings* have been used interchangeably.

A Neuroscientific Model of Mammalian Attachment

In neuroscientific models mammals, AT was described as the behavior of taking care of the offspring as soon as it is born, generally when she identifies the offspring as hers. In the seminal work of Jaak Panksepp (1998) on affective neuroscience, he found that, in rodents that exhibit AT, the maternal care provided by the females to the offspring seems to quiet them and induce them to sleep. On the other hand, when separated from the mother, the offspring shows great *bodily agitation*, with the *emission of ultrasonic vocalizations*, identified by the author as a sign of suffering, and concomitant activation of *the neural circuits of fear*. When they are reunited with the mother, and from the moment the care is resumed by her, the animals quiet down and stop emitting vocalizations of suffering, squinting their eyes. Early separation between the puppy and the mother, as well as the isolation between the puppy and its peers, results in permanent changes in the functioning of the HPA axis, in rats. In humans, it was found that these situations impair the individual's ability to cope with stress (Leuner et al. 2012).

According to Jaak Panksepp, caregiving behaviors would be activated by the female's neuro-affective brain circuits at the time of the birth of the litter, orchestrated by OT release (Panksepp et al. 1978; Panksepp 1998). In Panksepp's animal neuroscience model, mainly developed from studies with rats and mice, there is a detailed description of psycho-neuro-endocrine-affective circuits, substrates of the *basic emotional brain systems* that support AT: *SEEKING, CARE, FEAR,* and *PANIC*, which, in his theory, he chose to write in capital letters (Panksepp 1998).

The development and intensification of AT seems to occur in stressful circumstances, which promote in the puppy the search for a safe base. Stress, accompanied by the release of corticosterone, also seems to facilitate the formation of couples; a recurring association between the activation of the HPA axis and the expression of social behaviors was also noted (Suchecki et al. 1995).

These findings seem to emphasize the evolutionary aspect of AT and social bonds in the preservation of the species, that is, in situations of stress, AT can function as an attenuating of suffering and helplessness, in terms of maternal care, or favoring the formation of groups and couples.

A Neuroscientific Model of Human Attachment

The development of technical equipment, tools, methods, reagents, and paradigm shifts from 1970 (Parker 2018) enabled the realization of unprecedented experiments and discoveries: data obtained in research with rodents, primates, and ungulates began to suggest that there would be an association between *oxytocin* (OT) and the development of AT. Currently, there are ample experimental indications that, at the neuroendocrine level, the exercise of AT via mothering is mainly mediated by the release of the neuropeptide OT, simultaneously in the mother and in the puppy or baby (Carter 1998; Uvnäs-Moberg 2003; Feldman et al. 2013).

In the current scenario of neurobiology, new relevant questions can be raised. One of them concerns the relationships between the models and roles presumably developed in the critical periods of AT formation, and memory, as we understand it today. We will not delve into this topic, but, given its importance, some questions will be registered here. How would these roles and models be stored in memory? As declarative, conscious memory, or as procedural memory and, therefore, unconscious? When it corresponds to very early situations, how will this memory be stored? Will it be a sensory memory? Pictorial? Encrypted? Auditive? What is the relationship between the affective valence (positive or negative) of these roles and their persistence in memory? Which brain structures would be activated by it?

In neuroscience, studies on memory had two initial paradigms in the 1970s: studies of patients with amnesia and studies with animal models. Since then, it has been created a database about visual, motor, tactile (texture discrimination) memory; memory and sleep; declarative and procedural memories; and also, about memory disruption, as well as a series of information about the brain structures involved in memory evocation, thanks to experiments done with functional neuroimaging (Squire 2004).

Here is how procedural memory, a good candidate for storing information about AT, is described by Squire (2009): "…Procedural knowledge referred primarily to skill-based information, where what has been learned is embedded in acquired procedures. The major distinction is between the capacity for conscious, declarative memory about facts and events and a collection of unconscious, nondeclarative memory abilities, such as skill learning and habit learning."

The author concludes: "… Declarative memory is *representational*. It provides a way of modeling the external world, and it is either true or false. Nondeclarative memory is neither true nor false. It is *dispositional* and is expressed through performance rather than recollection" (Squire and Wixted 2011).

Despite the relevance of these developments, there seems to be a lack of dialogue between neuroscientific data and the exercise of clinical practice. On the other hand, psychoanalysis has memories as its main raw material for work in a clinical setting, but it lacks the production of systemized data and/or to benefit from what neuroscience has already unveiled about the issues raised here. This interface could be more fruitful, especially because, as can be observed in Squire's quotes, some neuroscientific conceptualizations are quite close to psychoanalytic conceptions.

To conclude, let us mention some current trends in research, and future perspectives in the investigation of AT in contemporary neuroscience.

In current research, more than new questions, we observe a trend in the neuroscientific deepening of phenomena previously addressed in their behavioral dimension.

For example, based on the classification of attachment types made by Ainsworth in the Strange Situation Test, it was sought to investigate how the attachment pattern is formed: to what extent its phenotype would be influenced by genetic factors, by the parental bond, or by environmental factors? (Erkoreka et al. 2021). This type of investigation imposes many difficulties in identifying the variables, and, often, its results cannot be replicated.

Assuming that AT is a biological phenomenon with evolutionary bases, Oliveira and Fearon (2019) conducted a literature review, observing phenomena at the genetic, physiological, and brain levels, which highlights a situation of translational research, where the interaction of different domains of knowledge comes into consideration, in a complex and challenging study. We move from a more naive, intuitive approach, to the use of statistical tools, which guarantee better answers. In a meta-analysis, for example, it was found that the association between temperament and insecure attachment was almost null, but the association between temperament and resistant attachment was moderate (Groh et al. 2017).

From a methodological point of view, there were mentioned a diversity of methods and instruments, among which there are studies of *DNA and genetic polymorphisms; studies with images*, in an attempt to identify changes in brain structures in situations of experimental AT; and studies with *twins and adoption*.

There were studies to assess the relative weight of attachment determined by genes and the environment (Picardi et al. 2020). Feldman et al. (2013) suggest that behavioral and hormonal (OT) parental synchrony establishes the emerging neuropeptide organization that should influence subsequent social behavior. The functioning style of the oxytocinergic system would be transferred from the parents to the child through parental care patterns (epigenetic mechanism). Several bonds between parents and children are supported by the oxytocinergic system (Feldman et al. 2013). Evidence remains that the primary caregiver shapes and regulates physiological, neurophysiological, and psychological functioning (mind-body), therefore, along with the discovery that the attachment bond is governed by many factors: stress management and the transmission of the mothering culture to the next generation (Hofer 2006).

The great promise for the near future seems to be filled by *molecular biology* and *molecular genetics*, and this includes psychiatric diseases (Fonagy et al. 2023). The challenges remain in studies whose results cannot be replicated and clinical trials with groups that are too small to have statistical significance. In short, there is much work ahead.

References

Ainsworth MDS, Blehar MC, Wers E, Wall S (1978) Patterns of attachment: a psychological study of the strange situation Hillsdale: Erlbaum

Beltrame GB (2011) Bases neurobiológicas del apego: Revisión Temática. Cienc Psicol 5(1):69–81

Blatt SJ, Levy KN (2003) Attachment theory, psychoanalysis, personality development, and psychopathology. Psychoanal Inq 23(1):102–150

Bowlby J (1958) The nature of the child's tie to his mother. Int J Psychoanal. https://doi.org/10.4324/9780429475931-15. Corpus ID: 15539652. In: Bowlby J. Attachment, Basic Books. Copyright ©Tavistock Institute of Human Relations, 1969, 1982, p. 179

Bowlby J (1969) Attachment. In: Attachment and loss, vol I, 2nd edn. Basic Books, London

Bowlby J (1973) Attachment and loss. Vol. 2: Separation: anxiety and anger. New York, NY: Basic Books

Bowlby J (1980) Loss: sadness and depression. In: Attachment and loss, vol III. Basic Books, London

Bowlby J (1982) Loss: sadness and depression. Basic Books

Bowlby J (1984) Violence in the family as a disorder of the attachment and caregiving systems. Am J Psychoanal 44(1):9–27, 29–31

Bowlby J, Fry M, Ainsworth MDS, World Health Organization (1965) Childcare and the growth of love. Penguin Books, London

Bowlby J. (2002) Attachment, Basic Books. Copyright ©Tavistock Institute of Human Relations, 1969, 1982, p. 179

Bretherton I (1992) The origins of attachment theory: John Bowlby and Mary Ainsworth. Dev Psychol 28(5):759–775

Carter CS (1998) Neuroendocrine perspectives on social attachment and love. Psychoneuroendocrinology 23(8):779–818

Champagne FA, Curley JP (2009) Epigenetic mechanisms mediating the long-term effects of maternal care on development. Neurosci Biobehav Rev 33(4):593–600

Crews D (1998) The evolutionary antecedents to love. Psychoneuroendocrinology 23(8):751–764

Diamond D (1999) Attachment research and psychoanalysis. 1. Theoretical considerations. Psychoanal Inq (Special Issue), 19(4)

Erkoreka L, Zumarraga M, Arrue A, Zamalloa MI, Arnaiz A, Olivas O, Moreno-Calle T, Saez E, Garcia J, Marin E, Varela N, Gonzalez-Pinto A, Basterreche N (2021) Genetics of adult attachment: an updated review of the literature. World J Psychiatry 11(9):530–542

Feldman R, Gordon I, Influs M, Gutbir T, Ebstein RP (2013) Parental oxytocin and early caregiving jointly shape children's oxytocin response and social reciprocity. Neuropsychopharmacology 38(7):1154–1162

Fonagy P, Target M (1997) Attachment, and reflective function: their role in self-organization. Dev Psychopathol 9(4):679–700

Fonagy P, Gergely G, Jurist EL, Target M (2002) Affect regulation, mentalization, and the development of the self. Other Press, New York

Fonagy P, Luyten P, Allison E, Campbell C (2018) Reconciling psychoanalytic ideas with attachment theory. In: Shaver P, Cassidy J (eds) Handbook of attachment. Guilford Press, New York

Fonagy P, Campbell C, Luyten P (2023) Attachment, mentalizing and trauma: then (1992) and now (2022). Brain Sci 13(3):459

Groh AM, Narayan AJ, Bakermans-Kranenburg MJ, Roisman GI, Vaughn BE, Fearon RMP, van IJzendoorn MH. (2017) Attachment and temperament in the early life course: a meta-analytic review. Child Dev 88(3):770–795

Gubernick DJ (1981) Parent and infant attachment in mammals. In: Parental care in mammals. Springer US, Boston, pp 243–305

Gubernick DJ (ed) (2013) Parental care in mammals. Springer Science & Business Media, Berlin

Harlow HF (1959) Love in infant monkeys. Sci Am 200(6):68–74

References

Hess EH (1959) Imprinting, an effect of early experience, imprinting determines later social behavior in animals. Science 130(3368):133–141

Hess EH, Petrovich SB (2000) Ethology and attachment: a historical perspective. Behav Dev Bull 9(1):14–19

Hofer MA (2006) Psychobiological roots of early attachment. Curr Dir Psychol Sci 15(2):84–88

Hong YR, Park JS (2012) Impact of attachment, temperament, and parenting on human development. Korean J Pediatr 55(12):449–454

Kennedy JH, Kennedy CE (2004) Attachment theory: implications for school psychology. Psychol Sch 41(2):247–259

Klein M (2013) On identification. In: Heimann P, Klein M, Kyrle-Money R (eds) New directions in psychoanalysis. The significance of infant conflict in the pattern of adult behaviour. Routledge, London, pp 309–345

Leuner B, Caponiti JM, Gould E (2012) Oxytocin stimulates adult neurogenesis even under conditions of stress and elevated glucocorticoids. Wiley Online Library, New Jersey

Main M, Solomon J (1986) Discovery of an insecure-disorganized/disoriented attachment pattern. In: Brazelton TB, Yogman MW (eds) Affective development in infancy. Ablex Publishing, New York, pp 95–124

Mason WA, Mendoza SP (1998) Generic aspects of primate attachment: Parents, offspring, and mates. Psychoneuroendocrinology 23(8):765–766

Mendoza SP, Mason WA (1997) Attachment relationships in New World primates. Ann N Y Acad Sci 807:203–209

Nelson EE, Panksepp J (1998 May) Brain substrates of infant-mother attachment: contributions of opioids, oxytocin, and norepinephrine. Neurosci Biobehav Rev 22(3):437–452

Oliveira P, Fearon P (2019) The biological bases of attachment. Adopt Foster 3(3):274–293

Pallini S, Barcaccia B (2014) A meeting of the minds: John Bowlby encounters Jean Piaget. Rev Gen Psychol 18(4):287–292

Panksepp J (1998) Affective neuroscience: the foundations of human and animal emotions. Oxford University Press, New York

Panksepp J, Herman B, Conner R, Bishop P, Scott JP (1978) The biology of social attachments: opioids alleviate separation distress. Biol Psychiatry 13(5):607–618

Parker D (2018) Kuhnian revolutions in neuroscience: the role of tool development. Biol Philos 33(3):17

Picardi A, Giuliani E, Gigantesco A (2020) Genes, and environment in attachment. Neurosci Biobehav Rev 112:254–269

Porges SW (1998) Love: an emergent property of the mammalian autonomic nervous system. Psychoneuroendocrinology 23(8):837–861

Porges SW (2011) The polyvagal theory: neurophysiological foundations of emotions, attachment, communication, and self-regulation. Norton, New York

Saitovitch A, Lemaitre H, Rechtman E, Vinçon-Leite A, Calmon R, Grévent D, Dangouloff-Ros V, Brunelle F, Boddaert N, Zilbovicius M (2019) Neural and behavioral signature of human social perception. Sci Rep 9(1):9252

Schore AN (1994) Affect regulation and the origin of the self: the neurobiology of emotional development. Lawrence Erlbaum Associes Inc, Mahwah

Schore JR, Schore AN (2008) Modern attachment theory: the central role of affect regulation in development and treatment. Clin Soc Work J 36(1):9–20

Spitz RA (1945) Hospitalism; an inquiry into the genesis of psychiatric conditions in early childhood. Psychoanal Study Child 1:53–74

Squire LR (2004) Memory systems of the brain: a brief history and current perspective. Neurobiol Learn Mem 82(3):171–177

Squire LR (2009) Memory and brain systems: 1969–2009. J Neurosci 29(41):12711–12716

Squire LR, Wixted JT (2011) The cognitive neuroscience of human memory since H.M. Annu Rev Neurosci 34:259–288. https://doi.org/10.1146/annurev-neuro-061010-113720. PMID: 21456960; PMCID: PMC3192650

Sroufe LA (1983) Infant-caregiver attachment and patterns of adaption in preschool: the roots of maladaptation and competence. In: Perlmutter M (ed) Minnesota symposium in child psychology, vol 16. Erlbaum, Hillsdale, pp 41–91

Steele M (2003) Attachment, actual experience, and mental representation. In: Green V (ed) Emotional developmental psychoanalysis, attachment theory and neuroscience: creating connections. Routledge, London

Suchecki D, Nelson DY, Van Oers H, Levine S (1995) Activation and inhibition of the hypothalamic-pituitary-adrenal axis of the neonatal rat: effects of maternal deprivation. Psychoneuroendocrinology 20(2):169–182

Trevharten C (2003) Stepping away from the mirror: pride and shame in adventures of companionship – reflections on the nature and emotional needs of infant intersubjectivy. In: Carter CS (ed) Attachment and bonding: a new synthesis. MIT Press, London

Uvnäs-Moberg K (2003) The oxytocin factor: tapping the hormone of calm, love, and healing. Da Capo Press, Cambridge

Uvnäs-Moberg K, Carter CS (1998) Is there a neurobiology of love? Proceedings of a conference. Stockholm, Sweden, August 28–31, 1996. Psychoneuroendocrinology 23(8):749–1013

Waters E, Weinfield NS, Hamilton CE (2000) The stability of attachment security from infancy to adolescence and early adulthood: general discussion. Child Dev 71(3):703–706

Winnicott DW (1948) Children's hostels in war and peace. Br J Med Psychol 21(Pt 3):175–180

Winnicott DW (1953) Transitional objects and transitional phenomena; a study of the first not-me possession. Int J Psychoanal 34(2):89–97

Winnicott DW (1962) Ego integration and child development. In: Winnicott DW (ed) The motional processes, and the facilitating environment: studies in the theory of emotional development. Karnac Books, London, pp 56–63

Winnicott D (1970) The mother-infant experience of mutuality. In: Anthony AE, Bender T (eds) Parenthood: its psychology and psychopathology. Little, Brown & Co, Boston

Winnicott DW (1984) Through pediatrics to psychoanalysis: collected papers. Routledge, London

Winnicott DW [1990] (2017) Human nature. Routledge. e-book

Winnicott DW (2005) Playing and reality. Routledge, London

Chapter 6
Brain Development

Introduction

To capture a *neuroscientific* view on phenomena such as *affect, attachment,* and *cognitive and emotional development*, it is necessary to delve into the *organs* and *brain structures* that give rise and physical support to them. Thus, in addition to the explicit behaviors corresponding to the phenomena, we can also highlight the entire corporal framework—nervous system, brain, hormones, and neurotransmitters—responsible for those phenomena production.

Embryonic and Fetal Development of the Human Brain

In the process of development of a human embryo, it is possible to detect the first traces of a very primitive *brain* as earlier as in the third week of gestation. It emerges from the *neural plate* and the *neural crests*, which are the precursor structures of the nervous system. However, for the brain to reach its full functioning, there will be a long and complex journey ahead.

Embryonic development, followed by fetal development, takes place through qualitative and quantitative changes that occur simultaneously, at morphogenetic, neurogenetic, genetic, neuroendocrine, environmental, cellular, and molecular levels (and this list is not exhaustive).

This developmental process involves complex steps, finely synchronized, which occur in an invariant sequence, within a genetic programming, with variations according to the species and sex-dependent restriction parameters (Stiles and

Jernigan 2010). They also depend on the presence of environmental stimuli necessary for gene expression[1] specific for each stage of this process (Lobo 2008).

However, embryonic, and, subsequently, fetal development, are not rigid processes. On the contrary, they are dynamic and consist of continuous interactions between the formation of basic structures and environmental stimuli (Zhang et al. 2013).

For successful implantation of the zygote into the uterine wall, for example, it must occur within the critical period called *implantation window* (approximately, between 6 and 10 days after ovulation), in the presence of specific hormonal conditions in the uterine wall so that the zygote can attach to it (Zhang et al. 2013).

Morphological Development of the Embryo

As a result of fertilization, the first cell of the new being, called *zygote*, emerges by the fusion of the hereditary genetic material (DNA) received from the parents. In its apparent simplicity, the zygote is a *totipotent stem cell* (Condic 2014), with potential[2] to produce every type of cell necessary for the development of the being. The zygote is the carrier of all the programming needed for the process of life, including the development of an efficient nervous system, specific to the species.

At cellular level, the development of the embryo is fast, and it becomes increasingly complex. After fertilization, several consecutive mitoses occur,[3] until a compact sphere of cells is formed, called *morula*. This happens about 3 days after fertilization, still in the *uterine tubes* (formerly called Fallopian tubes).

Between 3 and 5 days after fertilization, the morula begins to form a hollow space within itself, which is filled with fluid. It is then called a *blastula* or *blastocyst*.

At this point, the blastula has already reached the uterus, and is ready to be implanted in its wall, where it will be transformed in the *embryonic disk*, which, in turn, will develop three germ layers, which will give rise to all embryonic tissues. From this point, the embryo is called *gastrula*. The germ layers are called *endoderm, mesoderm, and ectoderm*.

The endoderm is the innermost layer and will give rise to the epithelium of the digestive system, the respiratory system, and glands, among others.

The mesoderm is the intermediate layer that will give rise to the dermis, to connective and muscular tissues, to the circulatory and urogenital systems, and to the notochord.

---

[1] Gene expression is the generation of a phenotype, from the hereditary information contained in a specific gene, and some specific environmental stimuli. Phenotypes, in turn, are the observable characteristics resulting from the interaction between a genotype with the environment.

[2] The potency of a cell is its ability to generate new cells different from itself. The more primitive the cell, from embryologic point of view, the greater is its potency, that is, the greater the variety of lineage it will be able to produce.

[3] Mitosis is the division of a cell in half, forming two new cells, identical to the first one.

The ectoderm, which is the outermost layer, will form the skin and the hair follicles, the external surfaces of the eyes, teeth, mouth, and rectum, the pineal and pituitary glands and, finally, a rudimentary nervous system (Woodbury et al. 2002).

Formation of the Nervous System

The nervous system begins with the creation, by the mesoderm, of a structure called *notochord*, in the shape of a rod. It will serve as an axis for the formation of the *neural plate* (which, in turn, derives from the ectoderm), as an embryonic structure of the spinal column. The neural plate, then, folds over itself and closes, forming the *neural tube* and the *neural crests* (external to the tube), initiating the development of the nervous system. At this point, the embryo is called *neurula*.

Next, the neural tube begins to morphologically differentiate at its top, a bulb will be formed, which will give rise to the *brain*; the lower part will become the *spinal cord*. The *neural crests*, which are a cluster of cells near the neural tube, will give rise to the roots of the dorsal ganglia and to the connective tissue of the head and neck.

During the development of the brain, a nonlinear programmed process will take place, in which some structures will mature before others. There will also occur transient processes, in which neural circuits are formed, and then, are undone. At this point, one cannot speak of remodeling, as nothing yet has been modeled. It is a construction, in which many elements function simultaneously, sometimes synergically, sometimes competing between them.

Neuroendocrine Action in Brain Development

Within the neural tube, the process of creating neurons and glial cells is in full swing, with about 125,000 neural cells already present. In comparison, at birth, the brain will have about 100 billion neurons (Nash 1997), which were generated at an average rate of 250,000 per minute during gestation.

During the development of the nervous system, there will be an important interaction between neurons, glial cells, and the endocrine system. The latter is a fundamental factor in gene expression, and early hormonal *imprinting* in certain brain regions can occur, which can affect the plasticity of the nervous system for life (Garcia-Segura 2009).

Thyroid hormones, for example, play a fundamental role in regulating the morphological and functional maturation of all regions of the central nervous system (Helmreich and Tylee 2011). Growth hormones help determine the final number of neurons and glial cells (Garcia-Segura 2009).

Stress hormones (Devenport and Devenport 1985), sex hormones (McEwen and Milner 2017), and neuropeptides (Bakos et al. 2016) will also aid in this process of maturation and definition of brain and spinal cord structures.

Among environmental stimuli relevant to the baby's brain development, the nutritional and stress conditions of the pregnant woman stand out. When they are outside the norm, it can cause future negative consequences; but when they occur positively, they will pave the way for affect regulation (Garcia-Segura 2009).

Morphological Development of the Brain

In the third gestational week, three vesicles begin to develop at the top of the neural tube, dividing it into *prosencephalon, mesencephalon, and rhombencephalon*, which, a little later, will differentiate into two more vesicles (Fig. 6.1). At the same time, the prosencephalon begins to fold forward, and is still quite large in relation to the body, which also begins to fold. The five vesicles will give rise to all areas of the brain (Fig. 6.2).

Around the fourth gestational week,[4] the encephalon begins to develop; by the seventh week, it is possible to identify its main parts. Note that the encephalon is a primitive embryonic organ, but the brain will differentiate from it later.

The heart, the first system to develop in the embryo, begins to beat on the 23rd day of life (Abdulla et al. 2004).

Brain Development in the Fetal Period

At the end of the eighth gestational week, with the main organs and body systems developed, the embryo is now a *fetus*, until the end of gestation.

Thanks to neurogenesis and neural connectivity (synaptogenesis), the brain begins to form neural circuits. As a result, the baby, at birth, has already, at his disposal, a basic neural formation in operation, with which he will be able to start his life outside the maternal uterus.

In the fourth month of gestation, motor function will start, indicating the maturation of the brain, which is responsible for coordinating all body movements. In the fifth gestational month, the process of myelination of the spinal cord and the roots of the dorsal nerves begins. Between the fifth and sixth month, the brain has already six layers, and the neural circuits continue to mature.

In summary, during the fetal period the development, maturation, and refinement of the structures occur, which were outlined in the embryonic period.

---

[4] The first gestational week corresponds to the first day of the last menstruation, 2 weeks before fertilization.

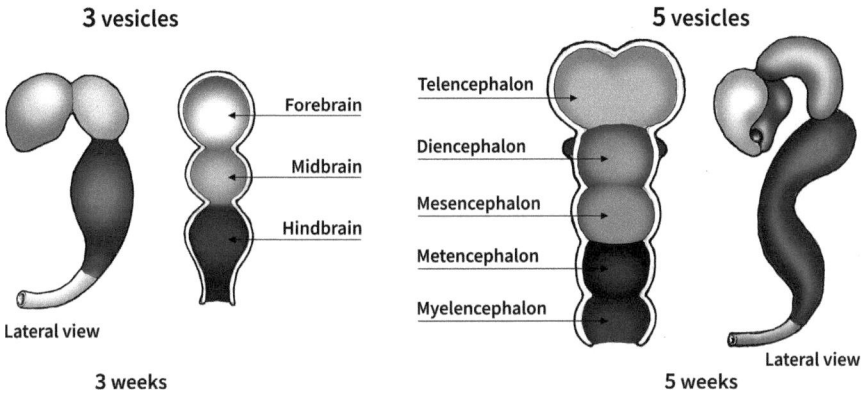

Fig. 6.1 Neural tube development: In the third gestational week, the neural tube will form three vesicles in its upper part, which will become five in the fifth week. At the same time, the tube begins to bend (see side views). (Illustration by Ciro Araujo)

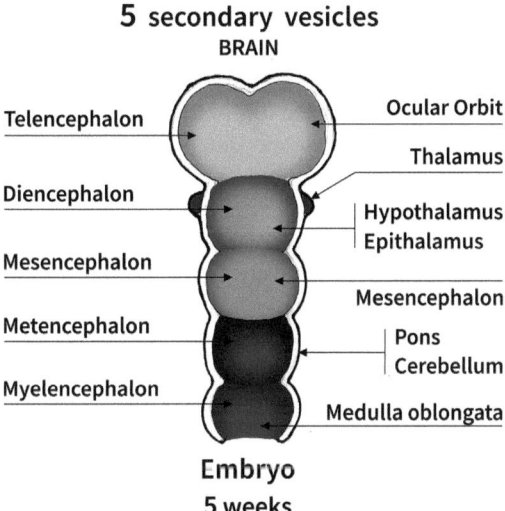

Fig. 6.2 Origin of the brain: Five vesicles will differentiate into the other brain structures. (Illustration by Ciro Araujo)

Neurogenesis

Neurogenesis consists in the birth and proliferation of *neurons* and *glial cells*, produced by *stem* and *progenitor cells*, a phenomenon that intensifies after the formation of the neural tube, in the embryonic period. After reaching a certain number, the juvenile neurons begin a migratory process, often in groups, until they reach their destination, where they will differentiate to assume distinct functions (Hatten 2002).

In the literature, there are many gaps, doubts, and contradictory findings still not clarified about neurogenesis. One of the reasons is that the research data on this subject comes mainly from animal studies. Most authors agree that, in humans, neurogenesis begins in the first half of the gestational period, followed by the migration of new neurons. However, there is no estimate from exactly when it begins and, even less, when it ends. Some scientists claim that neurogenesis ends during the gestational period (Malik et al. 2013), while others believe that it extends until the third trimester of pregnancy or even till the postnatal period (Sanai et al. 2011; Weiler 2018).

Other authors believe that, during human adolescence, there is a new period of intense neurogenesis in the dentate gyrus of the hippocampus, although some of these cells may die during the same period. Yet other authors believe that *there is no* neurogenesis in adolescence. Why are these findings so divergent? Among other factors, they are consequences of the use of different research methodologies, different sample sizes, access to more recent technologies, and lack of access to the human brain in vivo. We must emphasize that the findings about neurogenesis found in other mammals differ from those of humans. For example, in the former, we find neurogenesis throughout life; in humans, we do not.

In the adolescent brain, the survival of these cells seems to be linked to learning situations, and it is also vulnerable to stress, depression, and alcohol intake.

Concerning the possibility of neurogenesis during adult life, it is still a recent hypothesis, one received with optimism by scientists, as it suggests a possibility of regeneration of brain injuries and neurodegenerative diseases. However, it is still not clear exactly what this plasticity possibility of adult brains consists of if it indeed exists. Some researchers defend the presence of adult neurogenesis in certain brain niches linked to memory (Kumar et al. 2020), while others claim that neurogenesis would decline rapidly after birth (Sorrells et al. 2018).

However, how to explain mental modifications in adults? How to substantiate the result of psychotherapy, which implies the acquisition of learning and probably, the development of new neural circuits?

One of the authors suggested that, although there is no neurogenesis in adults, quiescent neurons with juvenile characteristics remain in the brain, which, in certain situations, could develop and assume new functions. Another hypothesis would be that these mental modifications could occur if there are some synaptic and neural circuit changes (Sorrells et al. 2018).

Data on adult neurogenesis are more abundant in studies with rodents, but the field of research with humans is beginning to become promising, thanks to the development of noninvasive methods and more powerful neuroimaging.

Cell Migration

Cell migration consists of the movement of cells, which occurs thanks to their own ability to generate movements toward certain directions and target locations which are phylogenetically preestablished.

During gestational development, cell migration usually occurs collectively, which is an advantage for the cohesion and organization of these groups. The transition from a static to a mobile state is linked to a program of morphological, structural, and molecular changes in the cell (Trepat et al. 2012).

Migration results in the formation of organized layers or columns of cells, which will differentiate into specific cells according to the location to which they are assigned (Rakic 1990).

Synaptogenesis and Neural Plasticity

Synaptogenesis, the process by which neurons connect to each other via synapse, constituting circuits or neural systems, occurs most intensely from mid-gestation until approximately 2 years of postnatal age (Abdulla et al. 2004).

Allied to neuroplasticity, synaptogenesis is the condition for the nervous system to be able to build, reorganize, or modify its neural connections in face of new situations or stimuli. This reorganization can result in the alteration, intensification, weakening, or modulation of the strength of neural circuits, and even the elimination of some connections (Taupin 2006).

Eric Kandel, an Austrian neuroscientist based in the United States and awarded the Nobel Prize in 2000, extensively studied the marine mollusk *Aplysia californica,* and demonstrated that, even being a very simple organism in neural terms, it shows a plasticity, albeit very limited (remember that an animal's learning must consider the specific limitations of the species in obtaining learning results). Kandel managed to train it to exhibit some behaviors, conditioned with an electric current. In this experiment, he was able to observe the phenomenon of *habituation*, in which the same stimulus, in this case, an electric shock, applied several times in the same place, ends up reducing the intensity of the response (Carew et al. 1981). In habituation, when a skill is often used, the organism reinforces the neural circuits and cellular mechanisms necessary to produce this behavior, understanding that it is important for survival or fitness of that being. With constant use and learning of behavior, the organism needs less and less effort and energy expenditure to repeat it (Thompson 2009).

Myelination

Myelination is a process that is part of the genetically programmed development of the nervous system, and its purpose is to increase the speed of the propagation of the neural message, in motor, sensory, and cognitive functions (Nave and Werner 2014). It begins in the second half of gestation and goes until the end of the second year of life, reappearing during adolescence and remaining until about the sixth decade of life (Benes et al. 1994). It also increases when the person has some intensive practice, for example, piano playing, especially in brain areas linked to memory (Fields 2015).

The *myelin* is formed by two types of glial cells: Schwann cells (originating from the neural crest and the peripheral nervous system) and oligodendrocytes (originating from the subventricular zone of the central nervous system). The process of myelination consists in the formation of a glial membrane that wraps around the axon several times, electrically isolating it in the section where this occurs (imagine a bare electric wire, around which an insulating tape is wrapped).

The sections of the axons isolated by the myelin sheath are interspersed with small segments without myelin, called Ranvier's nodes. This means that the only places where action potentials can occur are in these nodes, and therefore, the nerve impulse needs to jump over the myelinated section to the next node for its propagation. This myelinated axonal structure allows the increasing of its speed from twenty to a hundred times, compared to unmyelinated axons of the same diameter.

Neural Pruning

Neural pruning is a natural phenomenon which consists of the elimination of synapses and nerve cells at certain times in the development of the nervous system. Its function is to reduce cells, synapses, or groups of redundant, unnecessary, or excessive cells, which overload neural networks. This process, fundamental for the development of a healthy and adaptive brain, involves a competitive interaction between neural circuits, surviving the most efficient or functional ones for the being in its environment. During the first months after birth, synaptic pruning occurs after a period of axonal pruning. Then, there is a period of intense neurogenesis until the second year of life, followed by a long phase of neural pruning, as the baby adapts to its new condition in the world. This process can reduce the neural density of the cerebral cortex by about 50% (Sakai 2020).

Neural Pruning in Adolescence

A new period of neural pruning is genetically programmed to occur in *adolescence*, adding to other major changes underway in the brain. Thus, this pruning should be viewed in a transdisciplinary relation to other occurrences, such as the increasing myelination of axons, changes in the neurotransmitters *dopamine* and *serotonin*, and the onset of production of *gonadal hormones* and *morphological changes* in the brain that will influence the relationships between the cortical system and the subcortical system.

During neural pruning, approximately one-third of the dopaminergic receptors of the nucleus accumbens will be eliminated, with direct influence in the functioning of the *search and reward* systems and the psychological mechanisms of *motivation*, activated by dopamine. This means that the produced dopamine will have one-third fewer receptors to connect with (Spear 2013). The circuitry that links the ventral tegmental area, the nucleus accumbens and the prefrontal cortex (PFC), was called by Jaak Panksepp the *SEEKING system* (search system), which is linked to the behavior of exploring the environment, with the animal full of energy, motivation, and search for pleasurable stimuli. It is exactly this system that will undergo a decrease in area in adolescence, and it is important to investigate what behavioral tendencies are related to this phenomenon (Panksepp 1998).

Research on the adolescent brain is based in animal studies; therefore, it has limitations in translating it for humans. Magnetic resonance imaging (MRI) has also been used as an investigating tool. One of these studies found that neural pruning occurred in the cerebral cortex, more specifically in the prefrontal cortex, which, over time, was decreasing in size and losing neurons, dendrites, and synapses. The prefrontal cortex is the brain area that takes the longest to mature, and, among its assignments, is the inhibition and control of impulsivity, which, certainly, makes adolescents more prone to impulsive behaviors (Kim and Lee 2011).

While a decrease in gray matter begins, which will only be stabilized in adulthood, there is an increase in the amount of white matter, represented by the process of myelination of axons.

In addition to all this brain remodeling, adolescence also inaugurates a period of major hormonal changes, with the onset of production of gonadal hormones, associated with sexuality and readiness for sexual life (Juraska and Willing 2017). These brain changes, associated with hormonal ones and the onset of juvenile sexuality, with significant changes in the body, will bring a series of important behavioral consequences for adolescents: neuronal loss in the PFC, which is the part of the brain responsible for controlling impulses, and in the reward circuits. For this reason, the adolescent tends to seek much more exciting and intense stimuli to feel gratified. In the absence of this type of stimulus, it is possible that they may become depressed, demotivated, or bored, which will drive them to seek novelties that carry a greater amount of affective excitement. For this reason, the adolescent may be more prone to risky behaviors (automotive accidents, physical assaults, fights, early pregnancy).

If the adult's life circumstances undergo a major change, the brain modeling acquired in childhood will not always remain adaptive in adulthood, which could represent an additional difficulty to be faced.

References

Abdulla R, Blew GA, Holterman MJ (2004) Cardiovascular embryology. Pediatr Cardiol 25(3):191–200

Bakos J, Zatkova M, Bacova Z, Ostatnikova D (2016) The role of hypothalamic neuropeptides in neurogenesis and neuritogenesis. Neural Plast 2016:3276383

Benes FM, Turtle M, Khan Y, Farol P (1994) Myelination of a key relay zone in the hippocampal formation occurs in the human brain during childhood, adolescence, and adulthood. Arch Gen Psychiatry 51(6):477–484

Carew TJ, Walters ET, Kandel ER (1981) Classical conditioning in a simple withdrawal reflex in Aplysia californica. J Neurosci 1(12):1426–1437

Condic ML (2014) Totipotency: what it is and what it is not. Stem Cells Dev 23(8):796–812

Devenport LD, Devenport JA (1985) Adrenocortical hormones and brain growth: reversibility and differential sensitivity during development. Exp Neurol 90(1):44–52

Fields RD (2015) A new mechanism of nervous system plasticity: activity-dependent myelination. Nat Rev Neurosci 16(12):756–767

Garcia-Segura LM (2009) Hormones and brain plasticity. Oxford University Press, Oxford

Hatten ME (2002) New directions in neuronal migration. Science 297(5587):1660–1663

Helmreich DL, Tylee D (2011) Thyroid hormone regulation by stress and behavioral differences in adult male rats. Horm Behav 60(3):284–291

Juraska JM, Willing J (2017) Pubertal onset as a critical transition for neural development and cognition. Brain Res 1654(Pt B):87–94

Kim S, Lee D (2011) Prefrontal cortex and impulsive decision making. Biol Psychiatry. 69(12):1140–1146. https://doi.org/10.1016/j.biopsych.2010.07.005. Epub 2010 Aug 21. PMID: 20728878; PMCID: PMC2991430.Kumar A, Pareek V, Faiq MA

Kumar P, Kumari C, Singh HN, Ghosh SK (2020) Transcriptomic analysis of the signature of neurogenesis in human hippocampus suggests restricted progenitor cell progression post-childhood. IBRO Rep 9:224–232

Lobo I (2008) Environmental influences on gene expression. Nat Educ 1(1):39

Malik S, Vinukonda G, Vose LR, Diamond D, Bhimavarapu BB, Hu F, Zia MT, Hevner R, Zecevic N, Ballabh P (2013) Neurogenesis continues in the third trimester of pregnancy and is suppressed by premature birth. J Neurosci 33(2):411–423

McEwen BS, Milner TA (2017) Understanding the broad influence of sex hormones and sex differences in the brain. J Neurosci Res 95(1–2):24–39

Nash JM (1997) Fertile minds. Time 149(5):48–56

Nave KA, Werner HB (2014) Myelination of the nervous system: mechanisms and functions. Annu Rev Cell Dev Biol 30:503–533

Panksepp J (1998) Affective neuroscience: the foundations of human and animal emotions. Oxford University Press, Oxford

Rakic P (1990) Principles of neural cell migration. Experientia 46(9):882–891

Sakai J (2020) Core concept: how synaptic pruning shapes neural wiring during development and, possibly, in disease. Proc Natl Acad Sci USA 117(28):16096–16099

Sanai N, Nguyen T, Ihrie RA, Mirzadeh Z, Tsai HH, Wong M, Gupta N, Berger MS, Huang E, Garcia-Verdugo JM, Rowitch DH, Alvarez-Buylla A (2011) Corridors of migrating neurons in the human brain and their decline during infancy. Nature 478(7369):382–386

Sorrells SF, Paredes MF, Cebrian-Silla A, Sandoval K, Qi D, Kelley KW, James D, Mayer S, Chang J, Auguste KI, Chang EF, Gutierrez AJ, Kriegstein AR, Mathern GW, Oldham MC, Huang EJ, Garcia-Verdugo JM, Yang Z, Alvarez-Buylla A (2018) Human hippocampal neurogenesis drops sharply in children to undetectable levels in adults. Nature 555(7696):377–381

Spear LP (2013) Adolescent neurodevelopment. J Adolesc Health 52(2 Suppl 2):S7–S13

Stiles J, Jernigan TL (2010) The basics of brain development. Neuropsychol Rev 20(4):327–348

Taupin P (2006) Adult neurogenesis and neuroplasticity. Restor Neurol Neurosci 24(1):9–15

Thompson RF (2009) Habituation: a history. Neurobiol Learn Mem 92(2):127–134

Trepat X, Chen Z, Jacobson K (2012) Cell migration. Compr Physiol 2(4):2369–2392

Weiler N (2018) Birth of new neurons in the human hippocampus ends in childhood – adult "neurogenesis," observed in other species, appears not to occur in humans [Internet]. University of California San Francisco. [cited 2023 Dec 10]. Available from: https://www.ucsf.edu/news/2018/03/409986/birth-new-neurons-human-hippocampus-ends-childhood

Woodbury D, Reynolds K, Black IB (2002) Adult bone marrow stromal stem cells express germ-line, ectodermal, endodermal, and mesodermal genes prior to neurogenesis. J Neurosci Res 69(6):908–917

Zhang S, Lin H, Kong S, Wang S, Wang H, Wang H, Armant DR (2013) Physiological and molecular determinants of embryo implantation. Mol Asp Med 34(5):939–980

Chapter 7
Development and Affective Regulation

A saddled horse does not pass twice in the same place.
Gaucho saying

Windows of Opportunity

In Brazil, the above Gaucho saying refers to being able to take advantage of an important opportunity that presents itself. A good opportunity does not happen all the time: if a saddled horse passes near you, jump on its back, because you do not know when and if it will pass again; this concept of timing is suitable for talking about *windows of opportunity*.

The concept of windows of opportunity (or critical periods, or even, sensitive periods) is fundamental for understanding the ontogenetic development of the human being.

Windows of opportunity are genetically programmed temporal periods, in which new functions or brain structures are forming, and may come to express themselves as a phenotype, depending on whether they receive the right critical stimuli from the environment, and whether the human being can practice the new skill.

The windows have a certain period to open and close and represent the opportunity to optimally develop the new potentialities that the brain offers at that moment. If the opportunity is not taken advantage of, late learning will bring much more difficulty and energetic cost to the brain or may not even express itself anymore. If the learning process does not occur, or fails in the window of opportunity, the being will fail to develop a new behavior of its species and will become less adaptive to its world.

During the opening of a window of opportunity, optimal conditions of neural plasticity will occur, modulated by molecular, cellular, and hormonal conditions, for the acquisition of new skills (Ismail et al. 2017).

In ethology, there is a classic example of this phenomenon, related to the learning of song by songbirds. Based on data that indicate the assumption that, to develop its song, the bird would need to have contact with other birds of its species during a critical period, Marler (1970) observed that if isolated at birth, but exposed to the sound of its species' song recorded and played through a loudspeaker, the white-crowned sparrow could learn its own song if this occurred from day 10 to day 50 after birth; after that, it rejected both the sound of the song of birds of other species, and that of its own.

In neurobiological terms, and in a simplified way, this happens as follows: in the nucleus of each animal cell, there is a double helix of DNA, carrying its complete genetic code. At a critical moment, the helix splits in half, and one of the halves becomes a *messenger RNA (mRNA) strand*. The mRNA strand migrates from the nucleus to the cytoplasm of the cell and turns into a *chain of amino acids*. As it passes through a structure called *ribosome*, it turns into a *chain of polypeptides*. This chain folds several times over itself and turns into a *protein*. What will define the destiny of this protein is its form, because, depending on this and the stimuli from the environment, it will express a *phenotypic function*, in a process called *translational* (Edelman 1992).

Critical Periods in the Baby's Development

Prenatal Period

The prenatal period is, par excellence, a critical period for brain development, given the intense qualitative and quantitative modifications it will undergo from the third week of gestational life, when it is still undifferentiated from the neural tube (Sapunar et al. 2019). The earlier the stage of brain life, the more vulnerable it will be to toxic aggressions, and the more it will depend on critical stimuli to develop well. This is especially true for the first 3 gestational months, which is why doctors avoid administering drugs to the pregnant woman, or subject her to radiation during this period (Kumar and De Jesus 2023).

Maternal and Fetal Malnutrition

A fundamental factor for good brain development is that the baby receives adequate protein nutrition from its conception. Poor nutrition of the fetus can occur in situations that impair the passage of nutrients through the placenta, or when it receives little blood irrigation. In addition, genetic alterations, malnutrition of the pregnant woman during pregnancy and even before conception, and diseases she may carry can negatively affect the nutrition of the fetus.

Inadequate or insufficient fetal nutrition negatively interferes with neural proliferation and cell differentiation, causing the reduction of dendritic branches, synaptic connectivity, and causing cell death. In addition, metabolic and neurotransmitter systems can be impaired. The impairment of development can be permanent, manifesting itself as minimal brain dysfunctions of learning and memory. However, after birth, if the baby is given a healthy diet, integrated with good environmental stimulation, these negative conditions can be improved or even be reversed (Morgane et al. 1993).

A study published by McGaughy et al. in 2014 compared phenotypes of malnourished children to those of rodents that were subjected to very restrictive diets. Both groups exhibited attention deficits in changes of *setting* and decreased metabolic activity in the prefrontal region (area linked to executive functions).

In observational research that compared two groups of Chinese, one of which had been gestated or born during the Great Chinese Famine (1959–1961),[1] cognitive impairments were observed in the group that survived the famine in the rural area. Those from the urban area did not show deficits, which may indicate that access to a more enriched and stimulating world in the urban individuals may have compensated for the impairments of malnutrition (He et al. 2018).

Another study on the Chinese survivors of the Great Chinese Famine detected a probability twice as high of the development of schizophrenia in children gestated or born at that time (Smil 1999). Similarly, Xu et al. (2009) confirmed the relationship between fetal malnutrition, oxidative stress, and schizophrenia in rats.

Prenatal Maternal Stress and Outcomes for the Baby

Another factor with an extremely negative load for the fetus is composed by situations of stress to which the pregnant woman may be subjected.

A broad review of the literature coordinated by Franke et al. (2017) on the deleterious effects of maternal stress on the fetus considered a wide variety of stressors in pregnant women, such as psychological suffering, negative events, disasters, anxiety, and psychiatric diseases. When this occurred, negative consequences were observed in the neurodevelopment of children in cognitive aspects, in the display of difficult temperament, and, later in life, in psychiatric disorders and cardiovascular diseases. These phenomena are *epigenetic, not genetic* (Morange 2002), that is, they result from the interaction of the developing child with their environment. In favorable environments, the children can flourish; in unfavorable ones, they wither.

Rakers et al. (2020) investigated the neurobiological pathways through which maternal stress would harm the fetus. The cortisol produced by the mother's body is an important transplacental influencer, but there would be other factors, such as the dysregulation of catecholamines (adrenaline, norepinephrine, dopamine), which

---

[1] The Great Chinese Famine was the result of a decision by the Chinese Party leadership that, eager for great industrial growth that would match them with more developed countries, ordered peasants to abandon food planting and turn to the development of metal artifacts. These artifacts, however, were not commercially successful, and when winter came, there was no food. As a result, 30 million Chinese died of starvation in 3 years (Smil 1999).

would reduce uterine perfusion; the cytokines (immune system cells), serotonin, oxidative stress, and damage to the maternal microbiota, which would be transferred to the fetus. The authors bet on a hypothesis of damage by multiple stress factors acting together more than on a single cause. Similarly, Liisa Hantsoo et al. (2019) associated to the University of Pennsylvania, USA, proposed a multicausal model of stress risk factors during fetal life. They emphasized inflammatory events and fetal vulnerabilities as mediators of risk between maternal psychosocial stress and neural or psychiatric outcomes in babies.

Postnatal Period

For the newborn, birth is a radical and abrupt change of paradigm.

Suddenly, for the baby, the containment provided by the walls of the uterus disappears, as well as the gentle visual and thermal stimuli, the constant feeding, the damping of external noises and impacts, the constant rhythm of the mother's heartbeat, her characteristic smell: all those stimuli that ensured the well-being and continuity of the fetus in the uterine environment.

And at birth, what happens to him? What goes on inside a newborn? For those who live with him, there will be periods of crying and agitation, and periods of calm and tranquility. But can we say that the baby feels? Feels pain, feels fear.

Considering the immaturity of his brain, when he cries, he probably feels *primary affects*, still *nameless*, which can have *great intensity* (neurons are still premature in the process of myelination) and are probably hedonically *negative*.

Graham et al. (2016) published research on the emergence of fear in babies at 6 months of age, presumably because they noticed greater reactivity in the *brain's amygdalae* at 6 months than at birth.

Undoubtedly, the baby does not yet have cognitive functions to identify fear or pain like an older child. However, when he comes into the world, his brain's amygdalae are already functioning. In fact, the *amygdalae* were already *formed during the embryonic period*, long before birth, around the third week after conception, being the first part of the human striate complex to differentiate (Humphrey 1968).

As the amygdaloid complex is responsible for regulating anxiety, aggression, fear conditioning, and emotional memory, perhaps we can attribute to the amygdalae an immediate primary affective functioning in this early period, as the baby is already carrying, at birth a series of basic neural and affective programs to continue its life outside the uterus (Müller and O'Rahilly 2006). It will be up to the mother, or primary caregiver, the task of bringing order to the baby's life. And the baby will greatly benefit from the time he can spend in peace by his mother's, enjoying her company and interacting with her; this will leave a record of love and happiness in the baby's brain, which could become the basis of his mental construction.

It is important to ensure a stable sensory condition for the baby at birth, as similar as possible to that of the uterine environment. It would be a solution of continuity for a being still very immature from the neurological point of view, facing such a complex world. As myelination is still in progress, motor reactions to noises or sudden movements cause exaggerated spasmodic reflexes in the baby.

Like all living beings, the baby will develop its own rhythm, initially governed by its mother. Perhaps she will allow her baby to discover its own rhythm, or perhaps she will impose a rhythm on it as she read somewhere, which can be more, or less rigid, for eating, sleeping, changing diapers, bathing. At the beginning of life, the more stable and predictable this rhythm is, without exaggeration, the more the baby will consider the world a friendly place, without unpleasant jolts.

The baby's internal rhythm, as advocated by *chronobiology*, coordinated by the circadian light/dark cycle, will probably only be established later (Menna-Barreto and Wey 2007).

In general, mother and baby establish their rhythms in a synchronized way, which is not always easy for the mother.

Those who conduct psychotherapeutic clinical work and know how to listen to their patients will have certainly noticed that the earliest events leave marks on people. I remember, at this moment, reports where the mother had to distance herself from her baby for a while, due to illness or travel. These separations leave marks, which will be evoked much later by the now adult, in the form of discomfort, anxiety, pain. To take loving care of the baby, it is necessary to be close and synchronize with him.

The Baby and the Regulation of Affect

When talking about affect regulation, it is mandatory to mention Allan Schore, for his substantive contributions to this area.

Schore (1994) is an American neuroscientist from the University of San Francisco, California (UCLA), with multifaceted training in clinical neuropsychology, psychoanalysis, behavioral psychology, psychiatry, neurology, pediatrics, biology, and chemical biology. Integrating all these areas of knowledge, he became internationally known for having developed the *Theory of Affect Regulation* over the last 30 years, in which it stands out the *early regulatory role* of the *right cerebral hemisphere (RH)* (Schore and Schore 2017).

A regulated affective state is characterized by presenting a homeostatic balance (Fotopoulou et al. 2022), however dynamic, able to oscillate, but tending to rebalance afterward. If stressful situations are not too intense, or too prolonged, it is important that the baby learns to face stressful situations without disorganizing, through the development of *coping, self-regulation, and resilience* (Agorastos et al. 2018). This is achieved with the appropriate intervention of the mother. She is the baby's support and should learn how to use her feelings to know when he is hungry, when he wants to sleep, and when he is sick. Her role is to comfort the baby, slowly introducing the world to him, respecting his schedules, and protecting his rest. Initially, the baby must be treated according to its immaturity, and he should not be forced to grow before he is ready for it. I emphasize this because, currently, society treat babies like mini adults, making them go through very exhaustive and disruptive schedules, for which they are not yet neurobiologically prepared.

For Schore (2015), the process of affect regulation is intimately linked to the functioning of attachment mechanisms between the mother and her baby. Importantly, the Theory of Affect Regulation does not only refer to early childhood, but also about how a person can develop and blossom homeostatically to use the best of his potential, maintaining a healthy body and mind, developing an awareness about his world and about himself, and conditions for psychic survival. It also needs to develop a "theory of mind," being able to understand what the other is feeling and thinking, and what he is feeling and thinking at various moments of life, being able to, to predict the course of things in relation to another person. This is necessary, but not sufficient, for the process of humanization. For this, it must go a little further, and be able to see the other as if he were him, that is, to develop *empathy*.

It is also important to learn the language of its culture, and of diverse cultures he may get in touch with. What happens in her family is not always what happens in a friend's house, for example. But if he can learn at his own pace and rhythm, one thing at a time, rehearsing and making mistakes, it will have more opportunities to navigate in such a complex world.

Donald Winnicott speaks very sensitively about the baby's initial relationship with his world, and how novelties and unknown situations can be stressful for him (Winnicott 1984). The regulation of affect begins to form and has its most critical period in the first 2 years of life, but it is a project for life.

Schore, in the tradition of modern neuroscience, had intense scientific exchange with his fellows, which allowed him to add to his psychological knowledge, the infrastructure of neurobiology, neurodevelopment, endocrinology, and immunology. He also integrated brain, cognition, affects, and body, creating an extensive database of scientific information that allows for a deeper understanding of the processes of human ontogeny.

Right Cerebral Hemisphere

For Schore, affect regulation is neurobiologically anchored in the processing system constituted by the *right cerebral hemisphere* (RH). According to him, there is currently a substantial amount of research with imaging exams, which indicates that the RH matures well before the left hemisphere (LH), initiating a critical period of maturation which goes from the 25th gestational week until the beginning of the second year of life. Only then does the LH begin to have an accelerated growth. In the first 3 years of life, the RH is dominant in the baby's life, and its growth is not only due to its genetic constitution, but also to the influence of the typical emotional communications of attachment, which will literally influence the growth of neural tissues, brain hormones, and the child's brain. The importance of the RH is that it commands functions responsible for processing *nonverbal and unconscious* affective states.

After the development of the LH and the acquisition of language, the human being will go through conflicting processes about whether to use his speech to reveal or to hide himself, to be more oriented to the real world or to the construction of theories or fantasies. However, the RH will continue to be more important for the broader aspects of communication throughout life (Schore 2005).

Attachment and Right Hemisphere to Right Hemisphere Regulation in the Mother-Baby Dyad

The initial contacts between mother and baby occur through sensory stimuli, which begin to happen even before birth. The sense of smell was detected in fetuses after 30 weeks of gestation; it can discriminate odoriferous molecules in the amniotic fluid, reacting differently to foods ingested by the mother (Sarnat et al. 2017). There are also experimental indications that the fetus can identify the mother's smell (Vaglio 2009), her heartbeat (Porcaro et al. 2006), and the sound of her voice (DeCasper and Fifer 1980).

Therefore, at birth, there are already communication channels in action, so that the baby can have contact "with the outside," accompanied by a mother he already knows.

And why are these sensory aspects so important at the beginning of life? Because the LH is not yet ready to understand words, or rather, their cognitive content. Thus, attachment messages will basically run in his affective-emotional and unconscious sensory network, located in the subcortical regions of the brain.

When we examine the animal literature, we see that mammalian offspring, also respond sensorially to the presence of the mother, to the smell of milk, to her touch, and to the pressure of her body on them (Poindron et al. 2007).

In addition to the fact that RH favors a loving communication with the mother, its structure is also sensitive to the affective attention she dedicates to the baby, as this accelerates its growth. In other words, the RH and the cellular architecture of the cortex are built through the relationship with the primary caregiver. The brain is sculpted by these stimuli originated from initial relationships (Schore 2005).

There is another important brain structure called *orbitofrontal cortex* (*OFC*), which is more developed in the right hemisphere than in the left and is in the frontal region of the brain, above the eye orbits. At the end of the first year of life, the orbital areas of the frontal lobes will enter a critical period of growth, which will last until the middle of the second year. The OFC operates at the highest level of emotional behavior control, and Schore considers that it coincides with what Bowlby called the *attachment control system* (Schore 2018). For the proper development of this network, the processing of olfactory exchanges and glances between the mother and the baby will be fundamental (Robson 1967), and the baby will soon take the initiative to draw the mother's attention to itself, making noises with its mouth, or kicking vigorously when she enters its field of vision.

A study by Nitschke et al. (2004), which monitored the brains of mothers from fMRI images while they looked at photos of their babies, found that these mothers showed maximum activation in the right hemisphere OFC during this activity. The authors considered this to be evidence of the critical role of the representation of positive affect linked to attachment.

OT Production and Hedonic Sensory Stimuli

Initially, through animal studies, it was discovered that the production of OT is strongly associated with breastfeeding and care of the offspring.

Adequate mothering produces in the offspring behavioral responses that indicate *comfort*, characterized by *body quieting*, by *cessation of vocalizations*, by *closing of the eyes* and, eventually, by *falling asleep* (Panksepp 1998).

These responses, in turn, are mediated mainly by *contact*, by *touch*, by *pressure*, by *tepid temperature*, by *handling* and *holding*, all of them related to pleasant *sensory* stimuli.

On the other hand, when prematurely separated from the mother and placed in isolation, the animal reacts with *panic attacks*, with intense distress vocalizations *[distress]* and *motor agitation*, in opposition to the comfort response (Panksepp 1998). To those who work in the clinic, remember that patients who have panic attacks *should* be investigated for the experience of *abandonment*, *helplessness*, and *hopelessness*, and should receive, symbolically, the therapist's care to guarantee them a situation of *security and stability*.

Based on research with humans, Uvnäs-Moberg (2003) suggests that the activation of the *oxytocinergic system* should produce feelings of *calm, well-being, and comfort*, associated with a greater feeling of *trust* (Uvnäs-Moberg 2003). On the other hand, *stress states*, primarily organized by cortisol, play an important survival role and manifest themselves whenever a stimulus threatens homeostatic balance or the life of the being. These are situations usually experienced as *states of pain, fear, discomfort, or displeasure*, which quickly settle in the organism, generating intense, rapid, and automatic responses.

In the human adult, *slower and later cognitive reactions* can arise, aimed at restoring lost balance. On the other hand, when a person is in homeostatic balance, self-awareness can decrease, or even disappear—the body becomes "quiet," freeing the mind for other tasks. States of well-being, which should not be confused with moments of mania or euphoria, tend to be *subtle and discreet* in comparison to the exuberance and intensity of states of discomfort.

To close, the good part of OT is you do not need to ingest or inhale it to feel its effects of well-being. Your own body takes care of producing it. And this can be stimulated naturally: talking to someone you like and enjoy, having physical contact, hugging, kissing, listening to pleasant music, and getting a good massage. And this, of course, also applies to the baby.

References

Agorastos A, Pervanidou P, Chrousos GP, Kolaitis G (2018) Early life stress and trauma: developmental neuroendocrine aspects of prolonged stress system dysregulation. Hormones (Athens) 17(4):507–520

DeCasper AJ, Fifer WP (1980) Of human bonding: newborns prefer their mothers' voices. Science 208(4448):1174–1176

Edelman GM (1992) Bright air, brilliant fire: on the matter of the mind. Basic Books, New York

Fotopoulou A, von Mohr M, Krahé C (2022) Affective regulation through touch: homeostatic and allostatic mechanisms. Curr Opin Behav Sci 43:80–87

Franke K, Van den Bergh B, de Rooji S, Roseboom TJ (2017) Effects of prenatal stress on structural brain development and aging in humans. https://doi.org/10.1101/148916. License CC BY-NC-ND 4.0

Graham AM, Buss C, Rasmussen JM, Rudolph MD, Demeter DV, Gilmore JH, Styner M, Entringer S, Wadhwa PD, Fair DA (2016) Implications of newborn amygdala connectivity for fear and cognitive development at 6-months-of-age. Dev Cogn Neurosci 18:12–25

Hantsoo L, Kornfield S, Anguera MC, Epperson CN (2019) Inflammation: a proposed intermediary between maternal stress and offspring neuropsychiatric risk. Biol Psychiatry 85(2):97–106

He P, Liu L, Salas JMI, Guo C, Cheng Y, Chen G, Zheng X (2018) Prenatal malnutrition, and adult cognitive impairment: a natural experiment from the 1959–1961 Chinese famine. Br J Nutr 120(2):198–203

Humphrey T (1968) The development of the human amygdala during early embryonic life. J Comp Neurol 132(1):135–165

Ismail FY, Fatemi A, Johnston MV (2017) Cerebral plasticity: windows of opportunity in the developing brain. Eur J Paediatr Neurol 21(1):23–48

Kumar R, De Jesus O (2023) Radiation effects on the fetus. In: StatPearls [Internet]. StatPearls Publishing, Treasure Island (FL)

Marler P (1970) A comparative approach to vocal learning: song development in white-crowned sparrows. J Comp Physiol Psychol 71(2p2):1

McGaughy JA, Amaral AC, Rushmore RJ, Mokler DJ, Morgane PJ, Rosene DL, Galler JR (2014) Prenatal malnutrition leads to deficits in attentional set shifting and decreases metabolic activity in prefrontal subregions that control executive function. Dev Neurosci 36(6):532–541

Menna-Barreto L, Wey D (2007) Ontogênese do sistema de temporização: a construção e as reformas dos ritmos biológicos ao longo da vida humana. Psicol USP 18(2):133–153

Morange M (2002) The relations between genetics and epigenetics: a historical point of view. Ann N Y Acad Sci 981:50–60

Morgane PJ, Austin-LaFrance R, Bronzino J, Tonkiss J, Díaz-Cintra S, Cintra L, Kemper T, Galler JR (1993) Prenatal malnutrition, and development of the brain. Neurosci Biobehav Rev 17(1):91–128

Müller F, O'Rahilly R (2006) The amygdaloid complex and the medial and lateral ventricular eminences in staged human embryos. J Anat 208(5):547–564

Nitschke JB, Nelson EE, Rusch BD, Fox AS, Oakes TR, Davidson RJ (2004) Orbitofrontal cortex tracks positive mood in mothers viewing pictures of their newborn infants. NeuroImage 21(2):583–592

Panksepp J (1998) Affective neuroscience: the foundations of human and animal emotions. Oxford University Press, New York

Poindron P, Lévy F, Keller M (2007) Maternal responsiveness and maternal selectivity in domestic sheep and goats: the two facets of maternal attachment. Dev Psychobiol 49(1):54–70

Porcaro C, Zappasodi F, Barbati G, Salustri C, Pizzella V, Rossini PM, Tecchio F (2006) Fetal auditory responses to external sounds and mother's heartbeat: detection improved by Independent Component Analysis. Brain Res 1101(1):51–58

Rakers F, Rupprecht S, Dreiling M, Bergmeier C, Witte OW, Schwab M (2020) Transfer of maternal psychosocial stress to the fetus. Neurosci Biobehav Rev 117:185–197

Robson KS (1967) The role of eye-to-eye contact in maternal-infant attachment. J Child Psychol Psychiatry 8(1):13–25

Sapunar D, Arey LB, Rogers K (2019) Prenatal development | physiology. In: Encyclopædia Britannica [Internet]. [cited 2022 den 9]. Available from: https://www.britannica.com/science/prenatal-development

Sarnat HB, Flores-Sarnat L, Wei XC (2017) Olfactory development, part 1: function, from fetal perception to adult wine-tasting. J Child Neurol 32(6):566–578

Schore AN (1994) Affect regulation and the origin of the self: the neurobiology of emotional development. Psychology Press & Routledge Classic Editions

Schore AN (2005) Back to basics: attachment, affect regulation, and the developing right brain: linking developmental neuroscience to pediatrics. Pediatr Rev 26(6):204–217

Schore AN (2015) Affect regulation and the origin of the self: the neurobiology of emotional development. Routledge, London

Schore AN (2018) The experience-dependent maturation of an evaluative system in the cortex. In: Pribram KH (ed) Brain and values. Psychology Press, Equador, pp 337–358

Schore JR, Schore AN (2017) Regulation theory and affect regulation psychotherapy: a clinical primer. In: Miehls D, Applegate G (eds) Neurobiology and mental health clinical practice. Routledge, London, pp 44–61

Smil V (1999) China's great famine: 40 years later. BMJ 319(7225):1619–1621

Uvnäs-Moberg K (2003) The oxytocin factor: tapping the hormone of calm, love, and healing. Da Capo Press, Cambridge

Vaglio S (2009) Chemical communication and mother-infant recognition. Commun Integr Biol 2(3):279–281

Winnicott DW (1984) Through paediatrics to psychoanalysis: collected papers. Routledge, London

Xu MQ, Sun WS, Liu BX, Feng GY, Yu L, Yang L, He G, Sham P, Susser E, St Clair D, He L (2009) Prenatal malnutrition and adult schizophrenia: further evidence from the 1959–1961 Chinese famine. Schizophr Bull 35(3):568–576

Chapter 8
Therapeutic Choices

Introduction

There is usually a certain amount of confusion when talking about *psychiatry, psychoanalysis,* and *psychology. Affective neuroscience* is unfamiliar for lay people. *Psychopharmacology* is known as "the medications."

No wonder, in the tumultuous twentieth century, with the growing demands for psychic treatment, different therapeutic proposals emerged at the same moment, which ended up in an overlapping of them.

It is hard to make a choice when seeking treatment. One asks the professional: what *school* do you belong to? What is your *approach*?—as if the answer could help in making a choice. But there are hidden difficulties in this search. There is the fear that, by revealing their status of patient, people might think that he or she is not capable of solving their own problems. And that they might be *crazy*. Although prejudice is not as relentless as it once was, it still exists. *Madness* and *mental illness* are words that scare.

The French philosopher Michel Foucault, in his masterpiece *History of Madness*, clarifies that mental illness itself, as we understand it today, is a historical construction that goes beyond scientific practices, and that *madness*, with the connotation of *"mental illness,"* is a very recent reference in history (Foucault and Khalfa 2006).

Among those who seek treatment, there is a wide spectrum of cases that range from the very severe ones to those where people want to undergo therapy to better understand themselves and lead a fuller life; and each will need to be attended according to their level of expectation and complexity of their problem. Choosing a therapist is not just about choosing a line of work, but also about choosing someone you can trust and with who you are willing to share your intimacy with.

To help with these questions, we shall examine the most known treatment options and discuss their characteristics and estimated efficacy.

Psychiatry

At the end of the nineteenth and beginning of the twentieth century, *psychiatry* was dedicated to the confinement, in asylums, of people considered *mad* or *mentally ill*.

What it had to offer were water treatments (hot and cold baths, some prolonged and unpleasant) (Smith 2015); the surgical ones, which were exhausted in *lobotomies and brain ablations*, and the *electroconvulsive therapy* (ECT).

Conducted from 1935, lobotomies were surgical interventions which consisted in the removal of a piece of the brain, or of the lesion in a part of it; they were adopted for a certain time and then discontinued, as they were controversial and iatrogenic consequences. Occasionally, this treatment calmed down the patient, but also caused memory loss, damage to cognition, social and emotional skills, all irreversible, since, in surgery, the nervous tissues were damaged (Dibdin 2011).

ECT brought improvements to the patients, although its technique was still very primitive (Braslow and Marder 2019); patients were terrified at the possibility of being subjected to it, as it was administered without sedation (Suleman 2020).

In the first half of the twentieth century, parallel to the limited psychiatric repertoire above, it was beginning to grow an interest in a psychiatry based on science, which was making relevant discoveries about the *nervous system*. From there on, modern psychiatry changed its face many times since the discovery of the structure of neural networks (Bentivoglio 2023), at the end of the nineteenth century, and progressively became more scientific, biological, and technological. Before this era, science was, for the most part, guided by speculative premises (Barondes 1990).

Psychoanalysis

Psychoanalysis, developed by Sigmund Freud in Vienna, also at the end of the nineteenth and beginning of the twentieth century, grew in popularity in Europe, as the possibilities of the *talking cure* (Freud 1962) or *verbal therapy*, in which the patient was encouraged to express himself in words, were discovered—and this made all the difference in the treatment. The outpatient care was also a novelty, instead of the confinement in hospitals. And, no less important, the patient withdrew from the status of passive object of doctors and institutions.

Meanwhile, modern psychology was developing, perhaps in a less noisy way, working on a physiological and quantification line. In the United States, behaviorism achieved great popularity, but its scientific rigidity may have caused it to move away from the emotional world; thus, its therapeutic reach was more limited and superficial.

In opposition to this rigidity, psychoanalysis proposed a free association mode when the patient spoke, letting him lead the verbal flow, delving in aspects of his *Unconscious*, and, for this reason, it began to be called *depth psychology*.

Historical facts intervened in the development of psychoanalysis, causing ruptures, but also opening unexpected fronts of growth. Maybe the most relevant fact was that this (r)evolution in psychic treatments suffered the literal impact of two World Wars.

In Vienna, Europe, Freud was creating the nascent psychoanalysis. Around 1900, with a small group of disciples, Freud funded the *Wien Psychoanalytic Society*. He was concerned with preserving and strengthening the theoretical principles of psychoanalysis and in creating and spreading training institutes in other countries for psychoanalytic transmission. Not only the war came to disturb the psychoanalysis development, but the rise of Nazism started to create a dangerous environment for Jew psychoanalysts and unconventional treatments.

In the meantime, new associations were founded in other countries, of which we highlight those of Paris (France), London (England), and Budapest (Hungary) for their relevance, where psychoanalysis was becoming acculturated according to the conditions offered by each place. Regardless of the location, however, the psychoanalytic movement had, as a trademark, the engagement of its members in many fights over theory and power, which resulted in ruptures and schisms, with the consequent foundation of new centers.

The *Paris Psychoanalytic Society* was founded in 1926, without much repercussion. After the *Second World War*, it emerged in public institutions that offered psychiatric services, and there was practically no distinction between psychoanalysis and psychiatry. In other words, we have here a repetition of the dynamics in the United States, equally mobilized by war (Kapsambelis 2022).

In France, psychoanalysis attracted intellectuals and academics, apparently more interested in it than young doctors (de Mijolla 2010). In 1953, French psychoanalyst Jacques Lacan (1901–1981) broke up with the *International Psychoanalytical Association (IPA)*, proposing a new training system (Roudinesco 1990). From that moment, Lacanian psychoanalysis acquired a protagonism that lasted until 1970, when various dissident groups began to leave the École Freudienne de Paris. Even so, Lacan is still a reference in psychoanalysis in France.

An interesting and relevant observation for this chapter is that the psychoanalyst responsible for Lacan's didactic analysis, Rudolph Loewenstein (1898 1976), ended up migrating to the United States, where he developed an active career at the *New York Psychoanalytic Society*, and, together with two other Viennese psychoanalysts, Heinsz Hartmann (1894–1970) and Ernest Kris (1900–1957), created the *Ego psychology* (Hartmann et al. 1964).

What is amazing is that this brand of psychotherapy, which became dominant in the United States, was initiated by three European psychoanalysts, two of them originating from the Vienna school, closing a circle initiated by Freudian psychoanalysis, followed by Lacanian psychoanalysis, and ending by Ego psychology.

In the London group, which started in 1913, soon arose virulent theoretical disputes, led by the groups of Melanie Klein and Anna Freud (Nogueira-Vale 2003).

These disputes were transferred to North America, where they continued heated. In London, in addition to the Klein and Anna Freud groups, the Winnicott group stood out, studying the bonds between mother and baby, sensitized by the works of the Bowlby on attachment in the early years of life.

The political situation created by the war favored the emergence of dictatorial and totalitarian governments, who were not interested in being questioned by psychoanalysts or anyone; therefore, in many places, the psychoanalytic training institutes were prohibited from operating (Covelli s/d n.d.).

The *Hungarian Psychoanalytical Society,* in Budapest, was dismantled in consequence of these political interventions and of Nazism, interrupting a flourishing psychoanalytical work in progress, which took a decade to restart. The country was also geopolitically shattered, having lost two-thirds of its territory in the destruction of the Austro-Hungarian Empire (Mészáros 1998). The events in Europe had consequences for the development of psychoanalysis in the United States.

At the beginning of the Second World War, the American Army subjected all the recruits to a psychiatric evaluation, with the elimination of 1.1 million which were considered psychologically vulnerable, among a total of 15 million. However, of those considered fit and who were sent to the front, a large contingent returned to the country in pieces, with "battle fatigue" and horrible flashbacks indicative of *post-traumatic stress*. These soldiers needed urgent care, but there was not any civilian or professional preparation to assist so many people (Smith 2020). At one point, psychiatrist William Menninger, doctor of the Army whose family had a psychiatric hospital in Topeka, Kansas, took over the command of the Army's psychiatric service, *introducing psychoanalytic practices* in the care of soldiers (Menninger 2004). In fact, besides psychoanalysis, no other treatment was known that dealt with *trauma, distress,* and *panic*. This care provided to the American Army brought enormous visibility and growing prestige to psychoanalysis in the country (Ruffalo 2018).

But what would this *psychoanalytic psychiatry* be? In North America, a type of psychoanalysis with its own characteristics was emerging: an Americanized, medicalized, and popularized version of psychoanalysis, which deliberately led the public to confuse psychoanalysis and psychiatry, and not by chance. This was a strategy used by the *American Psychoanalytic Association* in 1946, chaired by Menninger himself, who thought that psychoanalysis in the United States needed to unify to establish itself professionally on American territory (Plant 2005).

Meanwhile, in Europe, the so-called *psychoanalytic diaspora* was taking place: due to the Nazi persecution (many European psychoanalysts were Jews) and with the prohibition of the practice of psychoanalysis, most psychoanalytic professionals abandoned their homeland and migrated to other countries. Many settled in the United States.

The psychoanalysts who came from Europe brought with them a psychoanalysis split by theoretical disputes. On the other hand, American psychoanalysis, in its almost entirety, differed from the psychoanalysis proposed by Freud, which was aimed at working with the truth and was geared toward a branch that came to be called *Ego psychology*, with a more conformist objective, of adapting the individual to the environment, thus losing the virulence and questioning of the original psychoanalysis (Benvenuto and Siniscalco 1996). Ego psychology spread and was practically the only branch of American psychoanalysis for two decades (Wallerstein 2002), and we realized, in a previous paragraph, who were the responsible for its creation (Leader 2021).

From the psychoanalysts who had come from Europe, many had already completed their training analyses—mandatory for taking patients in analysis, according to the rules of the *International Psychoanalytic Association*—but it is possible that many psychoanalysts had a heterodox training in America, which relied more on texts, and had a heterodox training analysis, due to the lack of qualified didactic analysts (those accredited to conduct training analyses). Otto Kernberg, for example, a prominent Austrian American psychiatrist and psychoanalyst, after escaping from Europe, went to Chile to do his training, before settling in the United States (Kernberg 2013).

In the 1960s, at the peak of its popularity in the United States, psychoanalysis greatly influenced American culture. Besides generating a large clinical clientele, it created a new concept of human sexuality and ended up having a moral and cultural impact, according to how each one interpreted them. Concepts about the Unconscious, slips of the tongue, resistance, for example, became known initially in the more elitist circles, and then became popular.

After a few decades, and in the face of new scientific discoveries, the popularity of psychoanalysis, according to some authors, was cooling down (Ruffalo 2019). Psychoanalysis still aroused much fascination, but, they said, it lacked a more scientific basis. Even so, several aspects of psychoanalysis had already been incorporated into American medical culture, and a whole generation of psychiatrists had undergone psychoanalytic training (Ruffalo 2018).

Psychopharmacology

In the mid-1950s, significant discoveries in the field of psychopharmacology revolutionized existing treatments for severe psychiatric cases of schizophrenia and depression that did not respond to existing therapies. Nathan S. Kline (1916 1983), an American psychiatrist and researcher interested in the biochemical and endocrinological aspects of mental illnesses, researched some drugs, aiming at developing *sedatives and antidepressants*. While working with one of these drugs in a study with schizophrenic patients, he observed an improvement in 70% of them, much higher than what had been achieved with other treatments so far.

The introduction of these drugs into clinical practice had an immediate effect of emptying psychiatric hospitals and reintegrating dysfunctional patients into society. Research, then, turned to patients who had severe depression, also obtaining positive effects.

At this time, thanks mainly to Kline's research on chlorpromazine, the first antipsychotic psychoactive substance was developed, as well as the first antidepressants, such as lithium and imipramine (Hillhouse and Porter 2015).

These events marked the beginning of American *psychopharmacology*, and psychiatry began to enter a more *biological* phase, although it was still used in conjunction with psychoanalysis (Ruffalo 2018).

However, psychotropic drugs offered a series of advantages to patients compared to psychotherapies: they acted on the neurotransmission of brain synapses, without pain and without lesions, and their use, after a few days of latency, was able to modulate the behavior and feelings of patients, allowing the reintegration of many of them into normal activities; therefore, they were cheaper and faster in effects than psychoanalysis, which could take years to bring results. Unfortunately, about 30–50% of patients *did not benefit from psychopharmacological treatment* (Al-Harbi 2012), a problem that persists today and constitutes a significant limitation for this type of treatment.

For the pharmaceutical industry, the appearance and popularization of psychotropic drugs represented an extraordinary source of revenue. In 1988, with the appearance and marketing of fluoxetine, commercially known as Prozac, another sales boom occurred, and, only in 1988, two and a half million medical prescriptions were written for it in the United States (Mukherjee 2012). In addition to their efficacy, psychoactive medications had their sales boosted by more aggressive and efficient marketing techniques.

The success of these medications led the pharmaceutical industry to work in the development of other *antidepressant and antipsychotic* medications, as well as *sedative, tranquilizers, and analgesic* drugs. All of these medications, each in their own way, had the same goal: to alleviate the pain and discomfort of those who were suffering. The target audience for these medications increased more and more.

In 1995, an opioid medication for pain called *oxycodone* (commercial name OxyContin) was launched in the American market. Its launching received heavy investment in advertisement to sellers and doctors and was a sales success. Initially prescribed for oncological pain, its use quickly expanded to other types of pain, and even to other types of discomfort. Unfortunately, its advertising and leaflets proved to be misleading, as they claimed that the medication *did not cause addiction*, and that it was *safe*, posing no health risks. In fact, it was *addictive*, stronger than morphine, and *lethal* in overdoses (Keefe 2021).

Soon, a legion of addicted people had formed, trying to get a prescription at any cost; pharmacists were selling under the counter, doctors were receiving bribes to provide receipts; a market of dealers were selling the drug illegally, and there were new deaths, until, finally, a group of users decided to start reporting these occurrences. Despite their effort, it was very difficult to be taken seriously by the press, and even by government regulatory agencies, including the FDA. Large financial interests were at stake.

Not only common people, but also famous artists and singers died from the unnecessary and excessive ingestion of OxyContin. The pharmaceutical industry that produced (and is still producing) it, became internationally billionaire (Keefe 2021). This situation persists to this day, with an epidemic difficult to manage. Here, we share with the reader some reflections on this era between wars, in which the

"American Way of Life" and North America as "the promised land" assumed a more defined and concrete form.

The "American Way of Life" is a utopia about America; a land where everyone can have equal opportunities and be successful, if they work hard; a land of freedom and happiness, the latter mainly centered on the possession of material goods, such as a car, house ownership, a TV, and several TVs. It is also an ideal of beauty, of harmony. What is not said, but underlies this idea, is that happiness must be pursued at any cost, and therefore, *pain, suffering, and frustration must be avoided*. These ideas are not very clear in the minds of citizens, but they actively operate unconsciously in their mental and emotional functioning.

Many patients reasoned: "Why do therapy if I can just take a medicine and solve my problem?" Drug treatment could have a much faster effect and was cheaper than therapy, besides preventing the patient from having to deal with emotional issues. As a result, many psychiatric, psychological, and psychoanalytic treatments were discontinued and replaced by the prescription of drugs (Winston et al. 2005). On the other hand, some professionals and scientists were proclaiming that psychoanalysis did not have a scientific basis, and this argument apparently decreased interest in psychoanalytic treatments.

Painkillers, sedatives, barbiturates, and other psychotropic drugs, thus, began to be part of the family pharmacy, because it is necessary to be happy. It is not enough to be well; one must be better. It is quite common to watch in the movies—when the protagonist suffers a blow or receives bad news, he immediately grabs a bottle with whiskey, conveniently placed in the scene, fills a glass, and swallows the alcoholic drink as if it were a saving medicine.

In 1993, shortly after Prozac entered the market, psychiatrist Peter Kramer wrote the bestseller *Listening to Prozac,* in which he suggests that Prozac and other antidepressants could be a type of "personality *enhancers*" and transform a boring person into someone more interesting. He even coined the term *"cosmetic psychopharmacology"* (Kramer 1993; Ruffalo 2020), a beautifier of personalities!

Affective Neuroscience

At the end of the twentieth century, great scientific progress had been achieved, not only in terms of neurobiological discoveries and techniques to study them, but also by the improvement of neuroimaging, which allowed to see the inside of the human body in a more detailed way, and in operation (functional neuroimaging). The identification of neural complexes and pathways, with their networks and transmitters, was fundamental to resize the conception of mental phenomenon.

The development of IT made the new discoveries accessible in real time, to anyone, anywhere on the planet. It allowed the storage of an increasing number of data, which could be manipulated each time faster, thanks to the development of new electronic devices.

But a mere database is not enough for a praxis, it is only an accumulation of information. A new integrated organization and dynamics had to be designed, as well as a new epistemology, as a basis for neuroscience. Yet, a crucial element was still missing for the full comprehension of what is a human being: the comprehension of *affective-emotional structures* and its dynamics in the brain.

This gap appeared from the beginning of the twentieth century, when *emotions and feelings* were banned from experimental studies, as researchers did not consider that affective facts were scientific, claiming that they were impossible to measure; others even rejected their existence (LeDoux 1996).

In 1992, however, the Estonian American neuroscientist and psychobiologist Jaak Panksepp opened a new possibility to reintegrate the *affects and emotions* into the understanding of mental phenomena and their psychopathologies, using for the first time the term, coined by him, of *"affective neuroscience"* (Panksepp 1998).

Panksepp's method of investigation, in which he invested a substantial part of his life, consisted in the meticulous and detailed identification, in the brains of mammals, of loci where primary affects were generated when certain neural networks were stimulated. For that purpose, he used electrical, hormonal, or environmental stimuli, observing that, whenever a certain *subcortical network* was stimulated, a primary emotion, and always the same one, manifested itself in the animal. Still, when he stimulated regions of the cortex, he did not obtain affective-emotional responses. In other words, the manifestation of affects originated in the *lower and more primitive parts of the brain*.

But what led Panksepp to this localizationist hypothesis?

Being a *Darwinian neuro-evolutionist* and therefore considering that the brain develops from the *bottom up*, he hypothesized that the lower a structure is situated in the brain, the most primitive, ancestral, and unconscious it is (Panksepp 2010). Panksepp also observed that affects were preserved in species when they were adaptive for their survival, and that mammalian brains were very similar to each other from a neuroanatomical, neurochemical, and functional point of view (Panksepp 2010). As it was not possible to conduct this type of investigation in humans because it was invasive, he hypothesized, and later confirmed, that the human brain had a subcortical neuroanatomy and functioning analogous to those of lower animals as far as the generation of affects was concerned. He then raised many hypotheses about the functioning of human beings by observing animals, which were later confirmed (Panksepp 1998).

In this comparative neuroscience methodology, the specificities of each species' brain must be considered. In humans, for instance, the neocortex gives rise to higher mental phenomena, typical of the species; however, Panksepp reminds us that the human affective mind (subcortical) functions in a way analogous to lower animals, as it has been preserved for millennia, nestled *within* the brain, and is associated to human culture (Panksepp and Northoff 2009).

Mora-Bermúdez et al. (2016), when comparatively studying the brains of humans and chimpanzees, noted that the cyto-architecture, cell type composition, and neurogenic expression programs of both had a remarkable similarity. At the level of the neocortex, responsible for the higher functions of the human brain, however, they found an increase in cortical neurons from fetal development, a fact corroborated by several scientists.

Panksepp also reminds us that the *primary affective processes* are fundamental for understanding the basic aspects of many psychological processes and for understanding psychiatric disorders. From neural mapping, Panksepp was able to identify seven primary emotions in the *subcortical brain area*: SEEKIG, RAGE, FEAR, LUST, CARE, PLAY, and GRIEF (Panksepp 1998, 2010).

Affective neuroscience has built an essential body of knowledge for the foundations of psychiatry, which would be more adequately called *biological psychiatry*. History shows us that there was a continuum of discoveries over time, with great protagonism of neurosciences, which ended up integrating psychiatry, psychology, and psychoanalysis, and even modifying their contours. This was something predictable. As new knowledge from various areas of knowledge dialogue with each other in a *transdisciplinary* way, their borders are expanded and a new knowledge emerges. The *transdisciplinarity* would be one of the main epistemological characteristics of affective neuroscience, besides being a current trend in sciences in general (Forum on Neuroscience and Nervous System Disorders 2015).

Likewise, psychoanalysis can only benefit from dialoguing with affective neuroscience, being able to base and direct many aspects of treatment, which will be addressed in this chapter.

Conclusion

It now should be clear to the reader that in the United States there was, in fact, a (intentional) mix between psychiatry and psychoanalysis.

Secondly, in the United States, we did not have the introduction of psychoanalysis, but rather of *several psychoanalyses*, of which the one known as *Ego psychology* was, until 1960, the dominant strand (Chodorow 2016). Note that the interpretive work of the Unconscious, the hallmark of Freudian psychoanalysis, is absent in Ego psychology. Parallel to most of the Ego psychoanalysis, there was a minority of European psychoanalysts in the United States, belonging to the schools of Balint, Winnicott Fairbarn, Melanie Klein, and Anna Freud, who worked in a more orthodox way (Benvenuto and Siniscalco 1996).

As Busiol (2013) explains, the *Ego psychoanalysis* is based on reason, causality, and coherence, and fits well into the American value system, as it relies on the adaptation of the subject to the environment. Freud's psychoanalysis, on the contrary, assumes that the action of the Unconscious (Id) is radically opposed to the conscious discourse of the Ego. The handling of this difference is what produces the critical and subversive power of psychoanalysis. The psychoanalysis of the Unconscious stings, bothers, and disturbs the patient.

Another characteristic of the acculturation of psychoanalysis in North America was the appearance of new forms of psychotherapy with a varied nomenclature: *analytically based psychotherapy, psychodynamic psychotherapy, Ego psychotherapy*, parallel to *cognitive psychotherapy* and *cognitive-behavioral therapy*.

Efficacy of the Various Therapeutic Modalities

After an epistemological overview of these overlapping fields, an important question arises: Which of these approaches would have therapeutic efficacy? How would they impact one's quality of life?

The American Psychological Association (APA) and the Harvard Medical School (HMS) inform on their websites that, in people with *depression*, both medication and psychotherapy are equally effective in terms of improvement, but, according to meta-analysis data, the use of both together yields *signifitly more effective results* than when used separately (De Jonghe et al. 2001, Cuijpers et al. 2009, 2014, 2015. APA 2017). In another meta-analysis, Kamenov et al. (2017) also confirm that the combination of psychotherapy and pharmacotherapy was more effective than just one of them in the treatment of depression, in addition to improving the quality of life of the patient.

The index for depression is eloquent, due to the prevalence of this pathology in the adult population of North America. According to Gallup (2023), adult Americans with a diagnosis of depression showed a prevalence of 29%, approximately 10% points above those obtained in 2015, and the highest percentage recorded since 2015. In contrast, the global prevalence of depression in adults is currently estimated at 5% (WHO 2023).

Regarding the types of psychotherapy mentioned on psychotherapy versus medications, an aspect should be highlighted: all the psychotherapies mentioned in studies have a *cognitive-behavioral affiliation*. Psychoanalysis is not in the list. Why not?

Several objections can be posed against psychoanalysis: it is not scientific; it does not have biological density; and it is not fit to participate in statistics. Moreover, it deals with *affections and feelings*, human aspects that have always been put apart in behavioral and cognitive approaches, because they were not measurable, they would not be reliable as variables, and so on (LeDoux 1996). Not least, psychoanalytic treatment is expensive and long term. There are also some variants of psychoanalysis, such as analytic-based psychotherapy and psychodynamic psychotherapy, also considered "depth psychologies," as they deal with unconscious aspects of the mind, work on transference aspects (relating to the relationship between analyst and patient as they occur in the sessions), even though this model accepts fewer weekly sessions than orthodox psychoanalysis, and a shorter period of duration (Novotney 2017).

Derived from cognitive psychology, there is also considerable prestige in the United States for the *cognitive-behavioral therapy*, described by the APA as an approach that assumes that cognitive, emotional, and behavioral aspects are interrelated (APA 2018). It sets goals and works with shorter periods than psychoanalysis.

Perhaps this refusal to work with the emotional world was made from superficial and ideological criteria; they also occurred at a time when the functioning of the brain was insufficiently known. But the prejudice persists today in many neuroscientists: that which is cognitive is much more valued than what is felt.

I raise this issue so that we can reflect on the relationship that exists, in our culture, between happiness and pain, and about how taking a psychotropic drug because one is unhappy ends up revealing how much discomfort needs to be suppressed, and not cared for. It is interesting that Kline himself, the creator of these especially important drugs, had psychoanalytic training, had psychoanalysts on his team, and clearly expressed his view on psychotherapy versus pharmacology. Ruffalo quotes Kline, in a text published on the blog *Mad in America*: "…the tranquilizing drugs should be used only for the treatment of those whose mental and emotional states disable them…," concluding: "mankind is perfectly capable of tranquilizing itself into oblivion" (Kline, 1957 *apud* Ruffalo 2017).

In summary, psychiatry and psychoanalysis have walked together, amalgamated, or in antagonistic poles, according to cultural circumstances and political interests, which also confuses the reader. As if that were not enough, psychiatry is increasingly integrated with psychopharmacology.

Affective neuroscience, by establishing a translational dialogue with psychoanalysis, helps to establish a neurobiological foundation for her.

The duration of psychoanalytic orientation treatments was positively associated with efficacy and improvement in mental health in long-term *follow-up*. At the beginning of the twenty-first century, it was found that even psychodynamic psychotherapy, of shorter duration, was equally efficient compared to cognitive-behavioral therapy in cases of anxiety and depression (Novotney 2017).

In the challenging task of understanding who the human being is in an integral way, it seems that concepts usually used by psychological science, such as *disease* and *sanity, homeostasis, allostasis, stress* and *allostatic load, neurosis, depression, or anxiety*, are far from encompassing all the dimensions of the being. How to fit, into these concepts, *suffering, pain, the broken heart, the lack of meaning in life, the destructive hatred*? Since antiquity, and after learning to survive the attacks of wild animals, hunger, and other threats to life, the human being had to face other challenges, now enriched by human culture and language, in an eternal search for *happiness*.

Next, in our final chapter, we will explore the possibilities of affect modulation psychotherapy, with the knowledge we have so far about oxytocin.

References

Al-Harbi KS (2012) Treatment-resistant depression: therapeutic trends, challenges, and future directions. Patient Prefer Adherence 6:369–388

American Psychological Association (2017) How do i choose between medication and therapy? American Psychological Association [Internet]. Available from: https://www.apa.org/ptsd-guideline/patients-and-families/medication-or-therapy

APA Dictionary of Psychology (2018) APA dictionary of psychology [Internet]. Apa.org. Available from: https://dictionary.apa.org/cognitive-behavior-therapy

Barondes SH (1990) The biological approach to psychiatry: history and prospects. J Neurosci 10(6):1707–1710

Bentivoglio M. Santiago Ramón y Cajal: life and discoveries. Nobel Prize.org. Nobel Prize Outreach AB 2023. Wed. 27 Dec 2023. https://www.nobelprize.org/prizes/medicine/1906/cajal/article/

Benvenuto S, Siniscalco R (1996) Otto Kernberg – psychoanalysis in America: interview. Eur J Psychoanal [Internet]. [cited 2023 Dez 19]. Available from: https://www.journal-psychoanalysis.eu/arasrticles/psychoanalysis-in-america

Braslow JT, Marder SR (2019) History of psychopharmacology. Annu Rev Clin Psychol 15:25–50

Busiol D (2013) How psychoanalysis was born in Europe [Internet]. [cited 2023 Oct 9]. Available from: https://drbusiol.com/index.php/tc/14-sample-data-articles/140-how-psychoanalysis-was-born-in-europe

Chodorow N (2016) Twentieth-century American psychoanalysis. In: Elliott A, Prager J (eds) The Routledge handbook of psychoanalysis in the social sciences and humanities. Routledge, London

Covelli G. Psychoanalysis after the war Britain, and Europe – History of psychoanalysis. https://www.academia.edu/36437987/Psychoanalysis_after_the_war_Britain_and_Europe

Cuijpers P, Dekker J, Hollon SD, Andersson G. (2009). Adding psychotherapy to pharmacotherapy in the treatment of depressive disorders in adults: a meta-analysis. J Clin Psychiatry; 70:1219–1229

Cuijpers P, Sijbrandij M, Koole SL, Andersson G, Beekman AT, Reynolds CF (2014). Adding psychotherapy to antidepressant medication in depression and anxiety disorders: a meta-analysis. World Psychiatry;13(1):56-67. https://doi.org/10.1002/wps.20089. PMID: 24497254; PMCID: PMC3918025.

Cuijpers P, De Wit L, Weitz E, Andersson G, JH Marcus, Huibers JH. (2015). The combination of psychotherapy and pharmacotherapy in the treatment of adult depression: a comprehensive meta-analysis Journal of evidencebased psychotherapies, 15(2):147–168

de Mijolla A (2010) Some distinctive features of the history of psychoanalysis in France. In: Birksted-Breen D, Flanders S, Gibeault A (eds) Reading French psychoanalysis. Routledge, London

de Jonghe F, Kool S, van Aalst G, Dekker J, Peen J. (2001). Combining psychotherapy and antidepressants in the treatment of depression. Journal of Affective Disorders, 64(2–3):217–229

Dibdin E (2011) The surprising history of the lobotomy [Internet]. Psych Central. [cited 2023 Dez 23]. Available from: https://psychcentral.com/blog/the-surprising-history-of-the-lobotomy

Forum on Neuroscience and Nervous System Disorders, Board on Health Sciences Policy, Institute of Medicine (2015) Developing a 21st century neuroscience workforce: workshop summary. National Academies Press (US), Washington (DC)

Foucault M, Khalfa J (eds) (2006) History of madness. Routledge, New York

Freud S (1962) The standard edition of the complete psychological works of Sigmund Freud: (1893–1895) Studies on hysteria. Hogarth Press, London

Hartmann H, Kris E, Loewenstein RM (1964) Papers on psychoanalytic psychology. Psychol Issues 4(14):1–206

Hillhouse TM, Porter JH (2015) A brief history of the development of antidepressant drugs: from monoamines to glutamate. Exp Clin Psychopharmacol 23(1):1–21

Kamenov K, Twomey C, Cabello M, Prina AM, Ayuso-Mateos JL (2017) The efficacy of psychotherapy, pharmacotherapy and their combination on functioning and quality of life in depression: a meta-analysis. Psychol Med 47(3):414–425

Kapsambelis V (2022) Psychoanalysis in the community in France. Int J Psychoanal 103(1):191–210

Keefe PR (2021) Empire of pain: the secret history of the Sackler dynasty. Doubleday, New York

Kernberg OF (2013) The development of a personal view of the psychoanalytic field. Psychoanal Dialogues 23(2):129–138

References

Kramer PD (1993) Listening to Prozac: a psychiatrist explores antidepressant drugs and the remaking of the self. Viking Press, New York

Leader D (2021) Lacan and the Americans. [Internet]. Eur J Psychoanal. [cited 2023 den 19]. Available from: https://www.journal-psychoanalysis.eu/articles/lacan-and-the-americans-d-leader/

LeDoux JE (1996) The emotional brain: the mysterious underpinnings of emotional life. Simon & Schuster, New York

Menninger WW (2004) Contributions of Dr. William C. Menninger to military psychiatry. Bull Menn Clin 68(4):277–296

Mészáros J (1998) The tragic success of European Psychoanalysis: "The Budapest School". Int Forum Psychoanal 7:207–214. Stockholm. ISSN 0803-

Mora-Bermúdez F, Badsha F, Kanton S, Camp JG, Vernot B, Köhler K, Voigt B, Okita K, Maricic T, He Z, Lachmann R, Pääbo S, Treutlein B, Huttner WB (2016) Differences and similarities between human and chimpanzee neural progenitors during cerebral cortex development. elife 5:e18683

Mukherjee S. Post Prozac Nation. The New York Times [Internet]. 2012 Apr 19 [cited 2023 Out 31]. Available from: https://www.nytimes.com/2012/04/22/magazine/the-science-and-history-of-treating-depression.html

Nogueira-Vale E (2003) Os rumos da psicanálise no Brasil: um estudo sobre a formação psicanalítica. Escuta, S. Paulo

Novotney A (2017) American Psychological Association. Psychoanalysis vs. psychodynamic therapy. https://www.apa.org [Internet]. [cited 2023 Out 31]. Available from: https://www.apa.org/monitor/2017/12/psychoanalysis-psychodynamic

Panksepp J (1998) Affective neuroscience: the foundations of human and animal emotions. Oxford University Press, New York

Panksepp J (2010) Affective neuroscience of the emotional BrainMind: evolutionary perspectives and implications for understanding depression. Dialogues Clin Neurosci 12(4):533–545

Panksepp J, Northoff G (2009) The trans-species core SELF: the emergence of active cultural and neuro-ecological agents through self-related processing within subcortical-cortical midline networks. Conscious Cogn 18(1):193–215

Plant RJ (2005) William Menninger and American psychoanalysis, 1946–48. Hist Psychiatry 16(62 Pt 2):181–202

Roudinesco E (1990) Jacques Lacan & Co.: a history of psychoanalysis in France, 1925–1985. University of Chicago Press, Chicago

Ruffalo ML (2017) Tranquilizing humanity into oblivion: a warning from Nathan S. Kline [Internet]. Mad In America. [cited 2023 Mar 17]. Available from: https://www.madinamerica.com/2017/11/tranquilizing-humanity-nathan-kline/

Ruffalo M (2018) The psychoanalytic tradition in American psychiatry: the basics [Internet]. Psychiatric Times. [cited 2023 June 20]. Available from: https://www.psychiatrictimes.com/view/psychoanalytic-tradition-american-psychiatry-basics

Ruffalo M (2019) The fall of psychoanalysis in American Psychiatry | Psychology Today [Internet]. www.psychologytoday.com. [cited 2023 June 20]. Available from: https://www.psychologytoday.com/us/blog/freud-fluoxetine/201912/the-fall-psychoanalysis-in-american-psychiatry

Ruffalo ML (2020) The story of Prozac: a landmark drug in psychiatry [Internet]. Psychology Today. [cited 2023 June 20]. Available from: https://www.psychologytoday.com/us/blog/freud-fluoxetine/202003/the-story-prozac-landmark-drug-in-psychiatry

Smith M (2015) The healing waters: the long history of using water to cure madness. [Internet]. Psychology Today. [cited 2023 June 9]. Available from: https://www.psychologytoday.com/us/blog/short-history-mental-health/201510/the-healing-waters

Smith M (2020) How the Second World War triggered concerns about mental illness in America. World War II and Mental Health | Psychology Today [Internet]. www.psychologytoday.com. [cited 2023 out 29]. Available from: https://www.psychologytoday.com/intl/blog/short-history-mental-health/202011/world-war-ii-and-mental-health

Suleman R (2020) A brief history of electroconvulsive therapy. Am J Psychiatry Resid J 16(1):6

Wallerstein RS (2002) The growth and transformation of American ego psychology. J Am Psychoanal Assoc 50(1):135–169

Winston A, Been H, Serby M (2005) Psychotherapy and psychopharmacology: different universes or an integrated future? J Psychother Integr 15(2):213–223

World Health Organization (2023) Depressive disorder (depression) [Internet]. World Health Organisation. [Internet]. [cited 2023 Dec 9]. Available from: https://www.who.int/news-room/fact-sheets/detail/depression

Chapter 9
A Conversation Between Affective Neuroscience and Psychoanalysis

Introduction

To close this book, we will present flashes from clinical cases, suggesting possibilities of clinical dialogue between affective neuroscience and psychoanalysis. We will use as parameters what was exposed in this text on *oxytocin, well-being, and affect regulation*, which are the central themes of this book. We will also introduce *neuropsychoanalysis* as a clinical tool for a deepening of some psychoanalytical principles.

The theoretical principles that underpin these suggestions were exposed throughout the former chapters, and to which the reader can return, if he/she wants to refresh his/her memory.

If you are not a psychoanalyst and work in another psychotherapeutic line, I still believe you can benefit from what we will comment here.

Initially, we will refer to some principles of *neuropsychoanalysis,* modality resulting from the dialogue of affective neuroscience with psychoanalysis, and to which we are aligned.

Neuropsychoanalysis began to take shape around 2000, in New York, by the initiative of an international multidisciplinary group, and, as far as we know, this was the first group to propose a dialogue between psychoanalysis and affective neuroscience (https://www.society@npsa-association.org). Many renowned psychoanalysts and neuroscientists joined the group, such as Arnold Pfeffer (psychoanalyst), Jaak Panksepp (neuroscientist and psychobiologist), Mark Solms (neurologist and psychoanalyst), Antonio Damásio (neurologist), Eric Kandel (neuroscientist), Vilaynur Ramachandran (neurologist), and Marianne Leuzinger-Bohleber (psychoanalyst). This group has become an international institution, with regular activities and annual congresses, a review, and centers in several countries, to develop neuropsychoanalytic concepts, presently under the name of *Neuropsychoanalysis Association* (N-PSA)

In its homepage, neuropsychoanalysis is defined as follows: *"... it is interested in the neurobiological underpinnings of how we act, think, and feel. As we begin to link brain activity with a psychoanalytic model of the mind, even at the deepest levels, a truly dynamic understanding can emerge."* (https://npsa-association.org/)

In their homepage, there are two guiding principles of neuropsychoanalysis:

Principle I: "*the brain is the organ that produces the mind. There is no possibility of pre-verbal or verbal, cognitive or emotional mental production without a brain to originate and sustain it.*"

Principle II: "*to understand more fully this production, the findings of neuroscience must be integrated at all levels of the mind.*"

This stitching between the neurobiological substrate and psychic phenomena, which occurs in an automatic and dynamic way according to the demands of the moment, is still poorly elucidated, and its development depends on future clinical trials and the ability of psychoanalysts and neuroscientists to keep dialoging and working together.

Starting from what we have exposed here, it is possible to make some clinical inferences. We will give some examples of clinical case excerpts, drawing attention to the neuroscientific and psychoanalytic aspects of it.

Clinical Case

The client was sitting on an armchair, talking about something he had done which could bring him negative consequences. Suddenly, he got excited, stood up, started pacing around the room, speaking loudly and gesticulating, without looking at the therapist. The patient's verbal language became confused, incongruent, and the emotional aspects, the motor agitation, stood out. At one point, he calmed down, sat back, and resumed his normal tone of voice.

What had happened at this point of the session at the patient's brain level?

To answer this questions, we will recall that, in neuro-evolutionary terms, the brain develops from bottom to top *[bottom up]*, with its *lower structures* in charge of generating and operating *primary processes*, the most primitive ones, and the higher structures generating more elaborate mental products. The higher they are in the brain, the more complex and cognitive processes they can generate (Panksepp and Biven 2012).

The primary processes generate *emotional feelings* that form an intrinsic value system, informing the person how they are doing in the world in terms of *survival*: whether they are at the moment, living in *comfort zones,* which favor life and well-being, or in *discomfort zones*, which threaten life, and are disagreeable (Panksepp and Biven 2012). These feelings, being so primitive, do not even have a name; they have not yet been symbolized/verbalized. They are unconscious, so that the person only feels *discomfort* or *well-being,* and, sometimes, they are so subtle that only another person, in this case, the analyst, detects them.

At the initial moment, when the patient was narrating what had happened, he was at a level of mental and verbal processing with reasonable control of cognitive content. Because he was performing this more cognitive task, we can infer that his cortex was being activated (the cortex is that wavy part on the outside of the brain, which has *associative sensory brain areas*, capable of generating *tertiary processes.*) Tertiary processes, integrally with primary and secondary processes, will generate what we know as *cognition* (Purves et al. 2001).

At a certain point in the patient's discourse, however, there occurred a disruption in this balance. Some stimuli, external or internal, must have triggered the change. The observable aspects of his behavior allow us to infer that his mental functioning has shifted downward, to the *subcortical part of the brain*, now recruiting lower structures, rooted more deeply in the brain, being, therefore, more emotional, more primitive, and more *unconscious*. We also noted that the patient manifested intense feelings of *displeasure* observable in his facial and bodily expressions. In other words, at that moment, the patient had entered in a *neuro-evolutionary regression*, in which the brain began to use the most primitive part of its structures, under the dominion of an unpleasant state, signifying that something threatened him.

We could have used the terms *regression* and *primitive* in their classic psychoanalytic sense. However, we are here referring to the functioning of *neurobiological areas* of the body/brain and its dynamics. These terms may even coincide with the psychoanalytic ones, but here we are using them referring to what happened at the brain/neural level, and consequently, in action, in behavior.

When there was a rupture in the patient's behavior, he showed a bodily agitation, an increase in voice volume, lost eye contact with me, and stood up. As he was *regressed*, he might not have much *awareness* of what he was saying/doing. Here the word "regressed" coincides, in its neuro-evolutionary sense, with the psychoanalytic sense. He is regressed and is not in a condition to function cognitively at that moment.

From data obtained from scientific studies in *affective neuroscience*, we know that some species of *mammalian offspring*, early in life, if separated from their mothers, go into despair. They become agitated, walk in circles, and emit vocalizations of suffering. When reunited with the mother, or if they are contained and sheltered in a warm place, they calm down, stop crying, and may even fall asleep. This important finding corresponds to the activation of the PANIC brain circuit, as identified by Panksepp, and is an experimental model of *helplessness* (and not of *fear*, as many think, which activates a different neural circuit) (Panksepp 1998).

This may be the most important and acute level of suffering in neuro-evolutionary terms, which is to find oneself alone, *helplessness*, something that, at this chronological age, poses a risk of death to the offspring. And that is why it hurts so much. The human baby, comparatively, presents similar manifestation, and others more specific to its species.

As we understood that the patient had gone through a *regressive moment*, more emotional, more primitive, and more unconscious, we waited for him to live through this moment to have greater contact with his emotions, and of which we remained as a witness. After he had vented them, I began to speak to him in a low and paused voice, calling him by his name, until he calmed down, could reconnect with us, look into our eyes, and talk about what had happened to him. In this moment, it possibly occurred what psychoanalysis calls *working through*.[1] (It is necessary that the analyst, while being supportive, allow the patient to live through this moment of pain, to have contact with it, and eventually, to be able to give it a name.)

Then, when addressing him, I began to call him by his name, so that he could regain his egoic condition. Only then, a reconnection was possible, a more mature eye and speech contact.

As we spoke calmly, and I called him by his name, the patient found himself supported by a *"maternal"* presence that took him out of anonymity, allowed him to calm down, and feel safe. Only then, after having regained the *state of well-being, of safety*, in which he should have produced *oxytocin* and had his affections and behavior *modulated* by contact with the therapist, it was time to deal with the *elaboration of integrated content* of what happened.

What would be the integrated content? It will emerge from the description we make to the patient of what we observed in him at an emotional-affective and behavioral level, and then, we pass the word to him. It will allow him to recover what belongs to him, at a more mature level, with which he can handle, and then *elaborate it* at a more cognitive level, proceeding to the work of psychoanalytic verbal content, which, in this case, referred to an abandoned child, while another, being a legitimate child, could have all the rights.

This would be a good example of the identification of neuro-evolutionary and psychoanalytic aspects in a session.

Basic Conditions for a Psychoanalysis

Psychoanalysis is a treatment that, according to its work contract, assumes that the client is *free* to accept it, for as long as he wants, being able to interrupt it at any time. (This condition is currently mandatory in any *Term of Agreement,* for any volunteer who wants to participate in a clinical trial).

In our proposal, to develop a healthy *therapeutic bond*, that is, to create conditions for the development of *attachment* (the vehicle through which psychoanalytic work will proceed), it is important that the patient has agreed to undergo therapy, or has sought it for free choice. It is also important that there is acceptance of the client for the indication of the psychoanalyst.

Even if the patient's choice is *imaginary*, since he does not yet know the therapist, or *transferential*, due to some information that he may have received about the

---

[1] Definition of *working through* in the *APA Dictionary of Psychology:* "In psychoanalysis, the process by which patients gradually overcome their resistance to the disclosure of unconscious material are brought face to face with the repressed feelings, threatening impulses, and internal conflicts at the root of their difficulties; and develop conscious ways to rebound from, resolve, or otherwise deal with these feelings, impulses and conflicts" (APA 2019).

therapist, the choice remains as a mark of a voluntary act of the client. The psychoanalyst also needs to listen to his internal availability and competence to meet the client's demand.

It cannot be emphasized enough that the analyst needs to have undergone, or be undergoing, a *training analysis,* to be able to attend the patient in a supportive and welcoming manner. If this precondition is not met, he will not be able to manage more tricky situations, or those in which his unconscious material may be confused with that of the patient.

The Setting in Psychoanalysis

The *setting* in psychoanalysis *is the psychoanalyst*, as the *setting* of the baby is the mother. I have attended patients in a pain clinic of a public institution, in Spartan facilities. I have attended in my office. And I have attended in transit, in emergency situations. Presently, I am attending online. And it seems to me that the connection with the patient is the figure; the rest is background. This connection arises when the analyst manages to establish a relationship of genuine interest in the patient and initiates an analytical listening.

Ardito and Rabellino (2011) mention that, from 1950 onward, there are works in the bibliography that discuss the therapeutic alliance between the professional and the client as a *primary factor in the outcome of treatment*. The authors say that, apparently, the quality of this alliance is a reliable predictor of positive clinical outcome, *regardless of the type of psychotherapy and outcome measures*. More recently, at a time when we have better tools to assess these relationships, a review with meta-analysis by Baier et al. (2020) confirmed this finding, which seemed to be independent of the theoretical line of psychotherapy, in 70.3% of the screened studies.

We propose that the analyst's interest in the patient should not be manifested in an exuberant or excessive way, but rather in a kind way. Oxytocin, which is produced in *attachment situations*, is not linked to euphoric emotions, but to calm ones. It is also especially important that the patient feels safe. This is achieved when the therapist manages to assess his level of suffering and, from this information, establishes enough weekly sessions to calm his anxiety. This number should be flexible, to adapt to different moments of the treatment. Many times, this maneuver was enough to regulate very anxious or depressed patients, without the need for medication. The sessions should be always in the same day and time; this brings a soothing feeling to a troubled soul.

If we think of maternal attachment as a *template*, we need to think about how to symbolically recreate it for the patient.

As we generally do not touch the patient in an analytical session—and *touch* is one of the most powerful stimuli to generate a situation of well-being and security—we could try to transform touch into *speech and gaze*. If the patient is very lost, he will need, at the beginning, of the therapist's gaze, so that they can constitute themselves as a subject.

For this reason, at the beginning of the treatment, we prefer to favor the gaze between the two, bringing back one of the earliest stimuli after birth, which is also fundamental for the formation of human attachment. This allows the psychoanalyst and the patient to have contact with the facial expression of the other, and to look each other in the eyes.

Here we already have the necessary ingredients for the creation of an attachment situation, with security and trust, for most of our patients. It is unprecedented to find the possibility of modulation of affect in these conditions.

It is implied that, in this approach, *lying on the couch* is not an immediate priority, and not even desirable. After the creation of some intimacy and positive emotions in the sessions, then we can think about the couch. This, in turn, is an important analytical tool: when moving from the face-to-face situation to the couch, the patient, when lying down and detaching from the analyst's gaze, can turn to themselves.

For many analysands, doing analysis on the couch brings a kind of status. I heard two people talking: *Are you doing psychoanalysis?* The other answered proudly: *Yes. On the couch.*

Resuming, it is preferable that, at the beginning of the work, the patient does not lie on the couch before establishing a more structured contact with the analyst. There will be plenty of time later for him/her to lie down, regress, and start a work of *free association*.

Physical Contact Between the Psychoanalyst and the Client

A short while ago, I said that, in our neuropsychoanalytic approach, the psychoanalyst's contact with the client is symbolic, that is, there should be no physical contact between them.

For many years, in psychoanalysis, this was an uncontested rule. Most psychoanalysts, including Freud, Lacan, and Winnicott, believed that the psychoanalyst should refrain from having physical contact with the patient, due to technical aspects related to the progress of the analysis. Freud himself used physical contact with his patients early in his work, but later criticized the practice, saying that there is intense transference in this situation. Hendel (2016) states that psychoanalytic theory has not yet endeavored to distinguish therapeutic touch from a sexual touch (it seems that psychoanalytic theory does not make this distinction, and in affective neuroscience, oxytocin, the hormone of human bonds, descends from an ancestral *sexual* hormone, vasotocin) (Panksepp 1998).

For Freud, as well as for Lacan, the *abstinence* to be maintained by the psychoanalyst is broader than physical touch; it meant that he should deny the patient the fulfillment of desire (a type of resistance) so that he could proceed with his analysis process. Even Winnicott, who valued the *holding* and the *handling* of the patient by the psychoanalyst, thought that contact through verbal interpretation could create a deeper sense of *holding* (Rodman 2004).

Some psychoanalysts took the rule of touch abstinence to the extreme, not even shaking hands with the patient. Others accepted social forms of greetings, which would be less eroticized, which also depends on local culture. In Brazil, it is common for women to kiss each other on both cheeks when meeting and saying goodbye, and a man and a woman who are somewhat close, too. In Anglo-Saxon countries, physical contact seems to be more restricted. Currently, even social touch is more inhibited, due to the campaigns against pedophilia and sexual harassment.

There are also some therapeutic lines, notably the *body therapies*, which not only accept this more social touch, but employ touch purposefully during treatment, for therapeutic purposes.

There are numerous body therapies difficult to submit to quantitative research, as they can be very heterogeneous in their proposals. It seems that it is especially difficult to discriminate certain variables, when it comes to therapies that use both the verbal instrument and physical touch simultaneously.

Some of them are more best known, such as Gestalt therapy, psychodrama, and Somatic Experience ©, this last one a kind of touch therapy used in the treatment of traumas and PTSD created by Peter Levine, an American doctor, psychotherapist, and biophysicist (see Ergos Institute for Somatic Education-https://www.somatic-experiencing.com/about-peter). There is also the Esalen Institute, founded in 1962 in California (USA), dedicated to body experiences and massage (https://www.esalen.org/abbestout).

The so-called *massage therapy*, developed in the United States mainly by Tiffany Field, although reporting improvements in depression, painful states, and stress, in vagal tone and immunity, confirmed by chemical changes in urine (Field 2016; Kerr et al. 2019), seems to be a *body approach directed at health*, without verbal elaboration work. In the evaluation of the critical literature, this type of therapy is considered adjunctive to other type(s) of therapy.

In the practice of psychoanalysis, if the psychoanalyst does not have an excessively rigid posture in his professional practice, he will probably find himself in challenging situations regarding physical touch.

Clinical Case

I was in my office, waiting for a patient. The door to the room was ajar, and I was reading a book. The patient, a 24-year-old young lady, coming from a dysfunctional family, rushes into the office, crying copiously, and throws herself on her knees besides me, with her arms around my waist. (What would you do?) She sobs for a few minutes, calms down, and returns to her chair, after which we can talk about what had happened to her. As her entrance was abrupt, I barely had time to think about how to manage the situation, but I hugged the patient back. I could not refuse this request for physical comfort, given her desperation. This event repeated once more, and there was never again any attempt at physical closeness, apart from the customary formal kisses on arrival and departure (this is the custom in Brazil).

About the Bond Between the Psychoanalyst and Their Client

At the beginning of a psychoanalytic relationship, there is usually fear from *both* parties—that the analytic partner becomes *mute*. The therapist becomes mute because he/she has learned that psychoanalysts should behave this way, so that the patient can establish an analytic field of his own. The patient is mute because he/she is terrified, as he is facing a situation he has never experienced before, and which is heavily contaminated by comments and myths he must have heard.

If the patient does not start the conversation, it is my opinion that the analyst should take the initiative to ask how he is, what is the reason for starting an analysis, so that he feels more comfortable. Besides it, the analyst is the more mature person in the relationship and is in the caregiver's role (the *"maternal"* situation).

A truthful and authentic relationship assumes that the patient can be himself in front of the analyst. This usually does not happen easily, nor naturally at first, as what is spoken in an analysis is different from the talk in social conversation, with its excesses and small hypocrisies; a psychoanalytic conversation is something to be built. For the patient to be himself implies not wearing masks, not needing to win the other's love, nor impress him. They do not even need to speak if the moment asks for silence.

The relationship developed between the two should be informal, but respectful; and the psychoanalyst should use his normal voice, not posed, or produced in an electronic or infantilized tone, and devoid of the use of technical terms when addressing the patient. None of these elements are superfluous; we are considering here that the human voice, with its authentic intonation and modulation, its melody, is essential for the formation and strengthening of a true bond between the two. And the model to be followed, you must have recognized, is that of *maternal attachment*, the matrix of affective relationships with the other(s). Attachment is an evolutionary phenomenon responsible for providing *security* to the baby and is essential for their *survival.* This loving maternal bond would be the desirable model for a *clinical setting*.

One of the rules of psychoanalysis is that the patient should tell the analyst whatever comes his mind. This is not easy at all, as it often goes against the principles of courtesy and social education; it is a language that needs to be developed. In addition, it is the patient who will decide how far they want to reveal themselves to the analyst, and at what pace. Notice how managing all of this, ethically and appropriately, is also not an easy task for the analyst.

The attachment model, as presented at the beginning, presupposes that the analyst can create a *safe, comfortable, and trustworthy* place where the patient can feel at ease, that is, feel well-being.

It is important that, if the patient, especially if they are very fragile, can have easy access to the analyst, through a secretary, answering machine, or cell phone. It is important that the interval between sessions is not too long, and the sessions are punctual, so that he does not feel lost.

As we are psychoanalysts and symbolic beings, our form of containment and welcoming should be symbolic most of the time, that is, preferably without physical contact between the two, but translated into ease of access by the patient to the therapist.

And what do we expect to happen, in neuropsychoanalytic terms? That a secure bond can be formed, which will automatically favor the release of endogenous opioid neurohormones, oxytocin, vasopressin, and norepinephrine, which mediate the creation of bonds between congeners.

Therefore, these principles will be valid, no matter what the therapeutic line is followed by the professional.

When Simple Contact with the Patient Has Therapeutic Value

When working with chronic pain patients in a public institution, I was selecting files for attendance, from a pile that the nurse had separated for me. I then started calling the patients, one by one, hoping that someone would answer. At one point, a woman answered the phone, and I scheduled a first psychological appointment for her.

Clinical Case

Like many people who migrate to SP, this woman came from the interior of northeastern Brazil, a poor and backward region, and had very few resources for her material survival. Her vocabulary was pitiful; although she wanted to tell more things and details, she lacked the words. She had four adopted teenage children and lived mainly on resources received from social services of the city hall.

The doctors at our institution were taking care of the medication part, but, even so, it was difficult to control the strong pains she felt all over her body. I talked to her, and I really felt moved by her problems and suffering. She slept sitting up, with her head lying on the table, on a pillow, because of the pain. In a contrasting way, despite her pains, she was tall and good-looking. She came to the hospital by bus from the outskirts, and it took her 2 hours to make the journey.

I felt limited, imagining what could I do for her, in face of such a miserable condition of her. She then started to come every week, and we would talk about everyday matters. One day she arrived, and said: *Doctor, you saved my life. How so? You saved my life. The day you called me, I could not take it anymore, I thought my life had no way out. Then you called me.* Her face lit up; she could not put anything else into words, but her face said it all. Eventually, I stopped seeing her, but she has my phone number. To this day, she sends me messages from time to time, and says she loves me, and that I saved her life. I have the impression that I really did save her, not me, but the person who was willing to attend and receive her. In this case, so wordless, what was therapeutic was my presence.

Final Remarks

This chapter is more authorial than the others and brings my own reading of affective neuroscience with psychoanalysis. My work with patients occurs from connections that I make with them, inspired by the moment. The work is anchored in theoretical studies, in my personal training, but the practice is created as I understand what is happening.

I invite the therapist reader to do the same.

References

American Psychological Association (APA) (2019) Working through. In: APA dictionary of psychology. American Psychological Association. https://dictionary.apa.org/working-through. Accessed 29 Jan 2024

Ardito RB, Rabellino D (2011) Therapeutic alliance and outcome of psychotherapy: historical excursus, measurements, and prospects for research. Front Psychol 2:270

Baier AL, Kline AC, Feeny NC (2020) Therapeutic alliance as a mediator of change: a systematic review and evaluation of research. Clin Psychol Rev 82:101921

Field T (2016) Massage therapy research review. Complement Ther Clin Pract 24:19–31

Hendel HJ (2016) Thoughts on the theoretical use of touch in psychotherapy [Internet]. www.linkedin.com. [cited 2023 July 10]. Available from: https://www.linkedin.com/pulse/thoughts-theoretical-use-touch-psychotherapy-hilary-jacobs-hendel

Kerr F, Wiechula R, Feo R, Schultz T, Kitson A (2019) Neurophysiology of human touch and eye gaze in therapeutic relationships and healing: a scoping review. JBI Database System Rev Implement Rep 17(2):209–247

Panksepp J (1998) Affective neuroscience: the foundations of human and animal emotions. Oxford University Press, New York

Panksepp J, Biven L (2012) The archaeology of mind: neuroevolutionary origins of human emotions. WW Norton & Company, New York

Purves D, Augustine GJ, Fitzpatrick D et al (eds) (2001) Neuroscience, 2nd edn. Sinauer Associates, Sunderland (MA). Neuroglial Cells. Available from: https://www.ncbi.nlm.nih.gov/books/NBK10869/

Rodman, FR. (2004) Winnicott: Life and Work, Da Capo Books

Index

A
Affective neuroscience, 9, 10, 13, 15, 26, 32, 33, 54, 59, 85, 91–93, 95, 99–108
Affective regulation, 4, 5, 45, 75–82
Affect regulation, 5, 68, 79, 80, 99
Attachment (AT), 3, 4, 14, 46, 51–61, 65, 80–82, 88, 102–104, 106

B
Body-mind, 30
Brain development, 3, 65–73, 76

C
Cell migration, 70–71
Cortical brain, 19, 22, 27, 28
Critical period, 3, 45, 52, 60, 66, 75–81

E
Epigenetics, 23, 51, 55, 61, 77

H
Health, 5, 16, 19, 40, 90, 95
Hypothalamus, 38–40, 42
Hypothalamus-pituitary-adrenal (HPA) axis, 10, 40, 46, 59

L
Left hemisphere (LH), 28, 38, 80
Libido, 47
Loving bond, 4, 51

M
Maternal care, 3, 4, 13, 41, 59
Myelination, 28, 68, 71–73, 78, 79

N
Nervous system (NS), 2, 11, 13, 19–33, 44, 56, 59, 65–67, 71, 72, 86
Neuroendocrine mechanisms, 67–68
Neuropsychoanalysis, 15, 99, 100

O
Oxytocin (OT), 1–5, 9–16, 37–47, 60, 95, 99, 102–104, 107
Oxytocinergic system, 41–43, 46, 61, 82

P
Peptide, 13
Pituitary, 12, 38–40, 67
Psychiatry, 12, 15, 79, 85–88, 90, 93, 95
Psychoanalysis, 10, 14, 15, 47, 52, 58, 60, 79, 85–91, 93–95, 99–108
Psychology, 5, 11, 14, 16, 53, 54, 58, 79, 85–88, 93, 94
Psychopharmacology, 85, 89–91, 95

R
Regulation, 3, 5, 14, 20, 23, 30, 38, 41, 42, 44–46, 57, 79–82
Right hemisphere, 28, 81, 82

S
Subcortical brain, 19, 93
Survival, 4, 5, 19–21, 23, 25, 26, 40, 42, 45, 53, 56, 59, 70, 71, 80, 82, 92, 100, 106, 107

W
Well-being, 3, 5, 13, 14, 19, 40, 57, 78, 82, 99, 100, 102, 103, 106
Window of opportunity, 75

MIX
Papier aus verantwortungsvollen Quellen
Paper from responsible sources
FSC® C105338

If you have any concerns about our products,
you can contact us on
ProductSafety@springernature.com

In case Publisher is established outside the EU,
the EU authorized representative is:
**Springer Nature Customer Service Center GmbH
Europaplatz 3, 69115 Heidelberg, Germany**

Printed by Libri Plureos GmbH
in Hamburg, Germany